EAI/Springer Innovations in Communication and Computing

Series Editor
Imrich Chlamtac, European Alliance for Innovation, Ghent, Belgium

The impact of information technologies is creating a new world yet not fully understood. The extent and speed of economic, life style and social changes already perceived in everyday life is hard to estimate without understanding the technological driving forces behind it. This series presents contributed volumes featuring the latest research and development in the various information engineering technologies that play a key role in this process. The range of topics, focusing primarily on communications and computing engineering include, but are not limited to, wireless networks; mobile communication; design and learning; gaming; interaction; e-health and pervasive healthcare; energy management; smart grids; internet of things; cognitive radio networks; computation; cloud computing; ubiquitous connectivity, and in mode general smart living, smart cities, Internet of Things and more. The series publishes a combination of expanded papers selected from hosted and sponsored European Alliance for Innovation (EAI) conferences that present cutting edge, global research as well as provide new perspectives on traditional related engineering fields. This content, complemented with open calls for contribution of book titles and individual chapters, together maintain Springer's and EAI's high standards of academic excellence. The audience for the books consists of researchers, industry professionals, advanced level students as well as practitioners in related fields of activity include information and communication specialists, security experts, economists, urban planners, doctors, and in general representatives in all those walks of life affected ad contributing to the information revolution.

Indexing: This series is indexed in Scopus, Ei Compendex, and zbMATH.

About EAI - EAI is a grassroots member organization initiated through cooperation between businesses, public, private and government organizations to address the global challenges of Europe's future competitiveness and link the European Research community with its counterparts around the globe. EAI reaches out to hundreds of thousands of individual subscribers on all continents and collaborates with an institutional member base including Fortune 500 companies, government organizations, and educational institutions, provide a free research and innovation platform. Through its open free membership model EAI promotes a new research and innovation culture based on collaboration, connectivity and recognition of excellence by community.

Lucia Knapčíková • Dragan Peraković
Editors

6th EAI International Conference on Management of Manufacturing Systems

Springer

EAI
RESEARCH MEETS INNOVATION

Editors
Lucia Knapčíková
Department of Industrial Engineering and Informatics, Faculty of Manufacturing Technologies
Technical Univ Kosice
Presov, Slovakia

Dragan Peraković
Faculty of Transport & Traffic Sciences
University of Zagreb
Zagreb, Croatia

ISSN 2522-8595 ISSN 2522-8609 (electronic)
EAI/Springer Innovations in Communication and Computing
ISBN 978-3-030-96313-2 ISBN 978-3-030-96314-9 (eBook)
https://doi.org/10.1007/978-3-030-96314-9

This Springer imprint is published by the registered company Springer Nature Switzerland AG
The registered company address is: Gewerbestrasse 11, 6330 Cham, Switzerland

Preface to the 6th EAI International Conference on Management of Manufacturing Systems (EAI MMS 2021)

The EAI MMS 2021—6th EAI International Conference on Management of Manufacturing Systems aims to bring together world-leading academics and practitioners from the fields of management, economics, infrastructure planning, and manufacturing. The unique combination of fields and disciplines focuses on Industry 4.0, the power of smarter information, and opportunities to create a bridge between science and practice. We are delighted to introduce the sixth edition of the 2021 European Alliance for Innovation (EAI) International Conference on Management of Manufacturing Systems. This conference has brought together researchers, developers, and practitioners from around the world. In the light of the latest knowledge and findings from scientific projects, the authors present actual R&D trends in the given field. This issue defines the state of the art in the area and explores related topics for future research.

The technical program of EAI MMS 2021 consisted of 18 full papers, including two invited papers in oral presentation sessions at the main conference tracks. The conference tracks were as follows: Track 1—Industry 4.0; Track 2—technological innovation, focusing on smart techniques/smart technology; and Track 3—sustainability of management and manufacturing systems. Aside from the high-quality technical paper presentations, the technical program also featured two keynote speeches, one invited talk, and two technical workshops. The two keynote speeches were of Dr. h. c. mult. Prof. Eng. MSc. Juraj Sinay, Dr Sc., with his keynote speech "Hydrogen—the energy medium of the future. National Hydrogen Strategy of the Slovak Republic," and Dr. Brij B. Gupta with his speech "Cyber Security and Critical Infrastructure Protection: Opportunities and Future Prospective."

It was also a great pleasure to work with such an excellent organizing committee team for their hard work organizing and supporting the conference. We are also grateful to the Conference Manager Aleksandra Sledziejowska for her support, all the authors who submitted their papers to the EAI MMS 2021 conference, and reviewers for their hard work.pagebreak

Our ambition was to establish communication channels and disseminate knowledge among professionals working in manufacturing and related institutions. We strongly believe that future cooperation will be as successful and stimulating as indicated by the contributions presented in this volume.

Presov, Slovakia
Zagreb, Croatia

Lucia Knapčíková
Dragan Peraković

Conference Organization

Steering Committee

Imrich Chlamtac
Bruno Kessler Professor, University of Trento, Italy
A/Prof. Lucia Knapčíková
Technical University of Kosice, Faculty of Manufacturing Technologies, Slovakia
Prof. Dragan Peraković, Ph.D.
University of Zagreb, Faculty of Transport and Traffic Science, Croatia
Assoc. Prof. Dr. Michal Balog
Technical University of Kosice, Faculty of Manufacturing Technologies, Slovakia

Organizing Committee

General Chairs

Assoc. Prof. Lucia Knapčíková
Technical University of Kosice, Faculty of Manufacturing Technologies, Slovakia
Prof. Dragan Peraković, Ph.D.
University of Zagreb, Faculty of Transport and Traffic Science, Croatia

General Co-Chairs

Assist. Prof. Dr. Marko Periša
University of Zagreb, Faculty of Transport and Traffic Science, Croatia
Assoc. Prof. Dr. Michal Balog
Technical University of Kosice, Faculty of Manufacturing Technologies, Slovakia
Assist. Prof. Dr. Annamária Behúnová
Technical University of Kosice, Faculty of Manufacturing Technologies, Slovakia

Technical Program Committee Chairs

Ivan Cvitić Ph.D.
University of Zagreb, Faculty of Transport and Traffic Science, Croatia
Dr. Jozef Husár
Technical University of Kosice, Faculty of Manufacturing Technologies, Slovakia

Web Chair

Dr. Jozef Husár
Technical University of Kosice, Faculty of Manufacturing Technologies, Slovakia

Publicity and Social Media Chairs

Assoc. Prof. Lucia Knapčíková
Technical University of Kosice, Faculty of Manufacturing Technologies, Slovakia
Prof. Dragan Peraković, Ph.D.
University of Zagreb, Faculty of Transport and Traffic Science, Croatia
Petra Zorić, mag. ing. traff.
University of Zagreb, Faculty of Transport and Traffic Science, Croatia

Workshops Chair

A/Prof. Dr. Michal Balog
Technical University of Kosice, Faculty of Manufacturing Technologies, Slovakia

Sponsorship and Exhibits Chair

A/Prof. Dr. Michal Balog
Technical University of Kosice, Faculty of Manufacturing Technologies, Slovakia

Publications Chairs

Assoc. Prof. Lucia Knapčíková
Technical University of Kosice, Faculty of Manufacturing Technologies, Slovakia
Prof. Dragan Peraković, Ph.D.
University of Zagreb, Faculty of Transport and Traffic Science, Croatia

Panels Chairs

Prof. Dr. Jan Pitel
Technical University of Kosice, Faculty of Manufacturing Technologies, Slovakia
Prof. Jerzy Wineczek
Czestochowa University of Technology, Faculty of Mechanical Engineering and Computer Science, Poland

Local Chair

Assoc. Prof. Lucia Knapčíková
Technical University of Kosice, Faculty of Manufacturing Technologies, Slovakia

Technical Program Committee

Technical Program Committee Chair

Ivan Cvitić Ph.D.
University of Zagreb, Faculty of Transport and Traffic Science, Croatia
Dr. Jozef Husár
Technical University of Kosice, Faculty of Manufacturing Technologies, Slovakia

Technical Program Committee Members

Dr. h. c. Prof. Dr. Jozef Zajac, Technical University of Kosice, Faculty of Manufacturing Technologies, Slovakia
A/Prof. Lucia Knapčíková, Technical University of Kosice, Faculty of Manufacturing Technologies, Slovakia
Prof. Dragan Peraković, Ph.D., University of Zagreb, Faculty of Transport and Traffic Science, Croatia
Assist. Prof. Dr. Annamária Behúnová, Technical University of Kosice, Faculty of Mining, Ecology, Process Control and Geotechnology, Slovakia
Assist. Prof. Dr. Marcel Behún, Technical University of Kosice, Faculty of Mining, Ecology, Process Control and Geotechnology, Slovakia
A/Prof. Dr. Michal Balog, Technical University of Kosice, Faculty of Manufacturing Technologies, Slovakia
Assoc. Prof. Milan Edl, Ph.D., University of West Bohemia, Faculty of Mechanical Engineering, Plzeň, Czech Republic
Prof. Dr. Janusz Grabara, Czestochowa University of Technology, Poland
Assist. Prof. Ivan Grgurević, Ph.D., University of Zagreb, Faculty of Transport and Traffic Science, Croatia
Assoc. Prof. Dr. Michal Hatala, Technical University of Kosice, Faculty of Manufacturing Technologies, Slovakia
Prof. Dr. Sergej Hloch, Technical University of Kosice, Faculty of Manufacturing Technologies, Slovakia
Prof. Dr. Michael Herzog, Technical University of Applied Sciences Wildau, Germany
Assoc. Prof. Dr. Vitalii Ivanov, Sumy State University, Ukraine
Assist. Prof. Gordana Jotanovic, Ph.D., Faculty of Transport and Traffic Engineering Doboj, University of East Sarajevo, BiH
Prof. Goran Marković, Ph.D., University of Belgrade, Faculty of Transport and Traffic Engineering, Serbia

Assist. Prof. Marko Periša, Ph.D., University of Zagreb, Faculty of Transport and Traffic Science, Croatia
Ivan Cvitić, Ph.D., University of Zagreb, Faculty of Transport and Traffic Science, Croatia
Prof. Dr. Ján Piteľ, Technical University of Kosice, Faculty of Manufacturing Technologies, Slovakia
Dr. Svetlana Radchenko, Technical University of Kosice, Faculty of Manufacturing Technologies, Slovakia
Dr. Jozef Husár, Technical University of Kosice, Faculty of Manufacturing Technologies, Slovakia
Prof. Dr. Alessandro Ruggiero, University of Salerno, Italy
Assoc. Prof. Dr. Oleg Zabolotnyi, Lutsk National Technical University, Ukraine
Prof. Dr. Jerzy Winczek, Czestochowa University of Technology, Poland
Assoc. Prof. Dr. Anna Saniuk, University of Zielona Gora, Poland
Prof. Dr. Dagmar Cagáňová, Slovak University of Technology, Bratislava, Slovakia
Dr. Natália Horňáková, Slovak University of Technology, Bratislava, Slovakia

Contents

Part I
Industry 4.0

Monitoring of an Industrial Process Based on Industry 4.0

Mauricio Xavier López Flores, Elvis Román López Flores, Francisco Javier Galora Silva, and Renato M. Toasa

1 Introduction

In today's environment of global competition, technological development, and innovation, companies, especially in manufacturing, are forced to constantly reconfigure their processes. The concept of Industry 4.0, based on a smart manufacturing approach, is part of this transformation, in which manufacturing and information technologies have been integrated to create innovative manufacturing, management, and business systems that optimize manufacturing processes, achieve greater flexibility and efficiency, and generate a value proposition for their customers, as well as respond in a timely manner to the needs of their market [1, 2].

The concept of Industry 4.0 refers to a government economic policy based on high-tech strategies, characterized by automation, digitization of processes, and use of electronic and information technologies in manufacturing [3]. It is also characterized by the personalization of production, provision of services, creation of value-added businesses, and capabilities of interaction and information exchange between humans and machines [4]. All this through the Internet of Things (IoT), allowing entire systems to interact with each other and with humans in real-time features that together with industrial automation are changing the way in which products are manufactured [5]. Contributing to the improvement of manufacturing and to companies having fully automated and interconnected processes that facil-

M. X. López Flores · E. R. López Flores · F. J. Galora Silva
Universidad Internacional de La Rioja en Ecuador, Quito, Ecuador
e-mail: mauricioxavier.lopez040@comunidadunir.net; elvis.lopez044@comunidadunir.net; francisco.galora049@comunidadunir.net

R. M. Toasa (✉)
Universidad Tecnológica Israel Quito, Quito, Ecuador
e-mail: rtoasa@uisrael.edu.ec

L. Knapčíková, D. Peraković (eds.), *6th EAI International Conference on Management of Manufacturing Systems*, EAI/Springer Innovations in Communication and Computing, https://doi.org/10.1007/978-3-030-96314-9_1

itate the flow of information, the decentralization of manufacturing, creation of new processes, decision-making, and focus on the development of competencies add value to organizations in such a way that generates not only greater innovation of products and processes, but also smart factories and other business models [6].

Hence, the integration of all activities of the company together with those who interact in the supply chain, suppliers, customers, and partners, within broad networks, is a core activity in organizations, and technology is the best means to design, create, and implement such environments that facilitate the exchange of information, products, and services, the exploitation of opportunities, and the creation of competitive advantages [7, 8].

Several emergent technologies as Industry 4.0 that is supported by the development of systems, the Internet of Things (IoT), and the Internet of People and services, in addition to other technologies such as additive manufacturing, 3D printing, reverse engineering, Big Data (massive data), and artificial intelligence, when working together, are generating transcendental changes not only in the manufacturing industry but also in consumer behavior and in the way of doing business [8]. Thus, at the same time, they favor the construction of capabilities that allow companies to adapt to market changes [9]. The technology of the coming years will be largely marked by the emergence of cyber-physical systems (CPS) [10], which will provide greater interconnection, speed, adaptability, and security to all types of products, services, and processes. All this will be vital for the growth of a company, generation of business, employment, and welfare of an increasingly demanding user/customer [11].

The literature shows that there are several papers that focus on the monitoring problem; initially Lennox et al. [12] developed a condition monitoring system for a fed-batch fermentation system; they mentioned that process monitoring and control have become important aspects of engineering; on the other hand, Severson et al. [13] developed an article that provided perspectives on progress in the development and implementation of process monitoring systems over the past 20 years, and discussed challenges in the field and opportunities for future research.

In Ecuador, a mobile system is being developed for real-time monitoring of the footwear manufacturing process [14]. The system is based on the andon, which monitors in real time the status of the industrial installation and the progress of the programming based on lean manufacturing techniques for notification, broadcasting, and visualization; another successful case of monitoring in Ecuador is shown in the work by León et al. [15], which describes the current situation of a wastewater treatment plant in a parish in Imbabura and a process that was formulated for the treatment and decontamination of water through a generalized monitoring system.

As stated in previous paragraphs, it is necessary to automate the monitoring of industrial processes through the development of industrial communication systems and platforms that allow adequate adaptation to the process, and their respective characteristics are of paramount importance. Therefore, and in accordance with the foregoing, this chapter, its analysis, study, and solution proposal respond to the need to formulate, implement, and have tools and resources that allow innovation in the current industry.

This work is divided into six sections including the Introduction. Section 2 presents the methodology, development of the proposal, and performance tests; Section 3 includes the results and their analysis; Section 4 determines conclusions and future work; and finally Sect. 5 includes bibliographical references.

2 Methodology and Platform Development

The development and execution of the proposal follow a logical order that includes the following subsections:

2.1 Architecture

The architecture scheme of the proposal is shown in Fig. 1. It shows a central broker that controls the communication, a client that maintains communication with the process controller and plant simulation, and another client hosted on a web page, from where process parameters can be visualized and controlled.

According to the requirement of the platform to be developed and taking into account the particularities of the process to be controlled, such as its virtual simulation in the Factory I/O software; the operation with variables of different types; and communication between platform, controller, and subsequent client/server/client

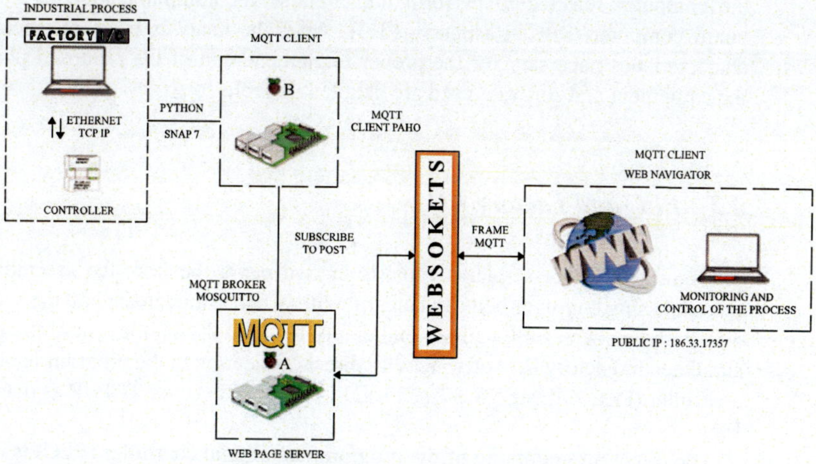

Fig. 1 Communication platform architecture

Table 1 Equipment and software used

Equipment	Detail	Description	Software/language	Link
PC	Wi-Fi access	Visualizes the process	Factory I/O	PLC
PLC	S7 1200 AC/DC 1214	Controls the process	TIA Portal V13	Factory I/O Raspberry Pi 3 with SNAP 7
Raspberry	Pi 3 B	Client MQTT	Language Python	PLC with SNAP 7 Server MQTT
Raspberry	Pi 3 B	Server MQTT	Language C	MQTT clients (WebSockets)
PC	Wi-Fi access	HMI Client MQTT	Web page	Server MQTT

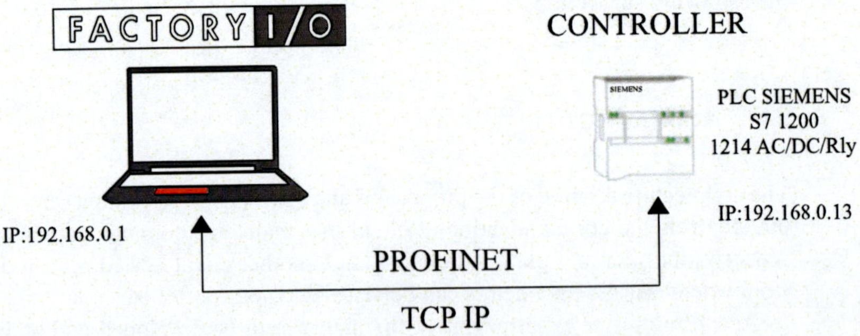

Fig. 2 Factory I/O and PLC connection

link, the controller to be used is a Siemens PLC S7 1200 1214 AC/DC/Rly, while the computer selected to perform the client/server communication is a single-board computer (SBC, Raspberry Pi 3). After the analysis of requirements and characteristics necessary for the proper implementation of the proposed platform, the equipment and software used are detailed in Table 1.

2.2 *PLC and Factory I/O*

The industrial process (plant) consists of two parts: the first, the assembly of a material using conveyor belts and an arm with X and Z movement and the second, a tank with level control for filling and emptying. The physical layout of the plant is simulated in Factory I/O software with direct connection to the programmable logic controller (PLC, SIEMENS S7 1200 1214 AC/DC/Rly) over TCP/IP as detailed in Fig. 2.

The direct programming of the programmable logic controller (PLC) is done in TIA Portal V13 software, in Ladder language. Among the most relevant instructions used are NORM_, SCALE_, mathematical instructions, counters, timers, system marks, and cycle marks, as well as the advanced PID_Compact block and its AutoTuning tool, among others.

2.3 PLC and Raspberry Pi

The data communication between the programmable logic controller (PLC) and the Raspberry to be used as client is established by means of the multiplatform programming tool Snap 7 allowing to obtain the data from the PLC Siemens S7 through the Profinet protocol (Profibus Ethernet). The communication is carried out by determining the classes, functions, and structure necessary for the correct operation of the reading and writing process in the specific memory addresses of the programmable logic controller used as client. The code syntax extract is detailed as follows:

```
class S71200():      # Determinación clase S71200
plc = S71200.S71200 ("192.168.0.13"): # Variable que llama a
clase s71200 y dirección IP del controlador
def getMem(self,mem,returnByte=False):     #Método para
lectura/escritura
```

Syntax for writing to the controller's real type memory and output:

```
plc.writeMem('QX0.0',True)    # Activar la salida 1 (Q0.0)
plc.writeMem('mw20',1200)     # Escribir el valor 1200 en la
memoria mw20
```

For reading of entries (IX0.0), marks (MX0.1), and whole words (MW20):

```
print plc.getMem('MX0.1')       # Leer el bit de memoria 2 (M0.1)
print plc.getMem('freal100')   # Leer el número real almacenado
en la memoria 100
```

These instructions are used to read and/or write the status of the input and output variables of the programmable logic controller used, thus allowing its remote control.

2.4 MQTT Protocol

The broker applied in the proposed platform is Mosquitto MQTT, a client-to-server publish/subscribe messaging transport protocol; it is lightweight, open, and applicable to machine-to-machine communication (M2M) and the Internet of Things (IoT), where network bandwidth limitations are not relevant. The complete structure made in the proposal, programs, software, and platform connection is shown in Fig. 3.

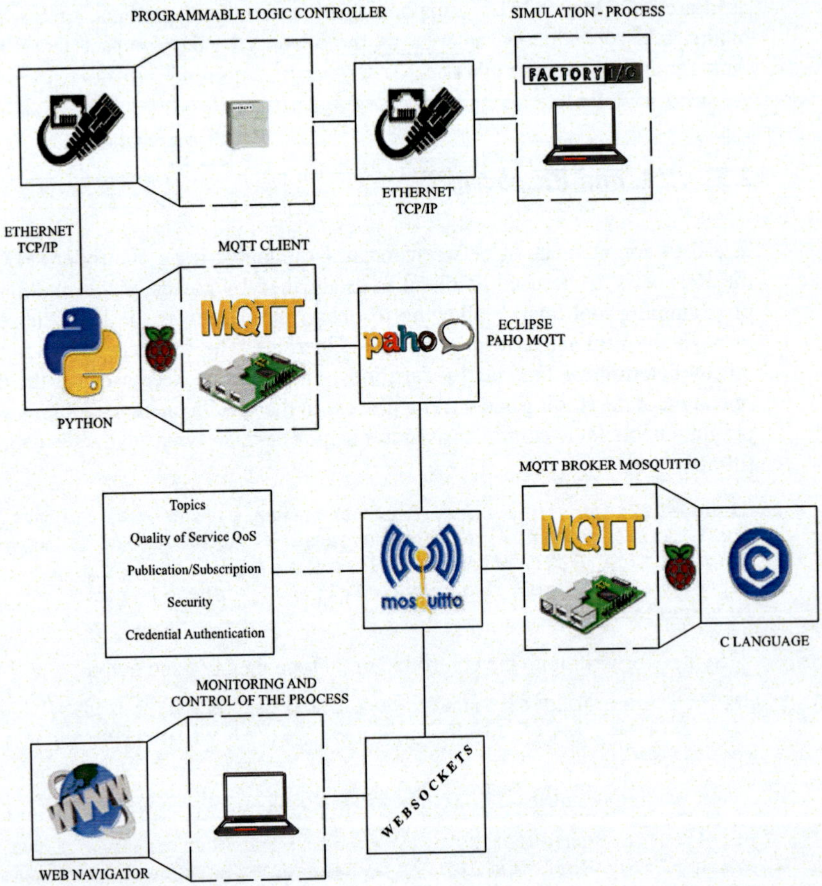

Fig. 3 Platform architecture

Validating the application and subsequent communication with browsers and web pages, the link of the MQTT protocol over WebSockets is determined. The Mosquitto MQTT broker presents communication ports, execution of native processes, and other default information, such as the operation on port 1883, in addition to identifying characteristics of new connections and client disconnections, warnings, errors, and other relevant references. Its operation is carried out in C language, with the objective of implementing MQTT clients, both publishers and subscribers, without major communication problems. These clients may have limited features and resources; however, by having and working under a TCP/IP stack both in the client and in the intermediary, the connection will be viable.

The MQTT protocol follows a star topology, with the central node Server Raspberry A as the broker; it is responsible for managing the network and transmit messages and also operates the two clients of the developed application; however, it can work with many clients. The string of characters necessary for the broker to operate and properly filter each message received is the Topic. Thus, if we refer to the value of the variable of the PLC Boolean type with the name "Start Mark," whether it is to publish or subscribe, the Topic required is *PLC/PLC1/ENSAMBLE/CONTROL/START.*

Through the MQTT publish/subscribe scheme used, a simple architecture is established that allows decoupling the client that sends a message from the one that receives it, the connection and communication controlled by the broker; filters the messages from the publishers; and distributes them appropriately to the subscribers. It is not necessary to create/declare the desired Topic before publishing or subscribing; the broker accepts each valid Topic without any prior initialization.

Command is executed from the client to send a message, and publication is composed of a Topic and the message to be sent: *mqttc.publish("TOPIC","True").*

The command executed from the client to subscribe to a Topic of interest consists of the topic and a quality of service (QoS) level, as follows: *mqttc.subscribe("TOPIC", 0).*

2.5 WebSockets

The application and development of the web page are supported by the use of the MQTT protocol over WebSockets, which in general allows sending and receiving messages directly from and to a browser, in addition to being able to operate all the features and functions of MQTT as usual.

WebSocket is based on TCP and as a protocol it provides bidirectional communication between a browser and a web server. Figure 4 shows the architecture used in the communication of the MQTT protocol over WebSockets.

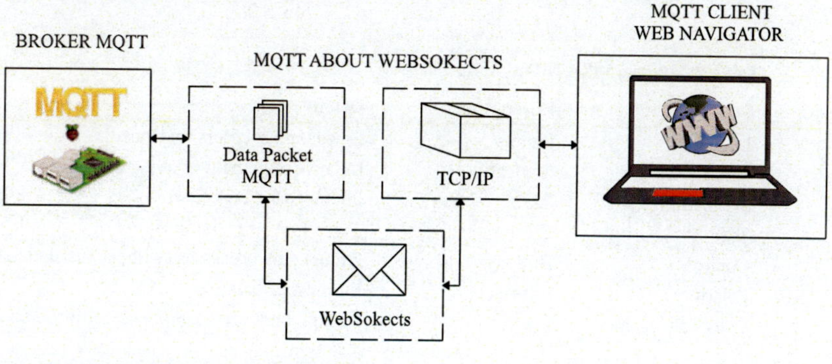

Fig. 4 MQTT architecture on WebSocket

The configuration performed allows confirming the ports open for communication by MQTT's Mosquitto broker; on WebSockets the port determined is 9001; in addition, connections, errors, disconnections, and other client alerts in the protocol are determined and updated, as well as their IP address.

2.6 Web Page: HMI

With the necessary configuration for the communication between the MQTT protocol on WebSockets, now the web page is developed for the corresponding visualization and control of the industrial process through the use and application of a LAMP server (Linux, Apache, MySQL, PHP), and programming of the base structure in HTML, configuration and improvement of the design and presentation in CSS, dynamism in JavaScript, complements and simulation integrated with the AJAX programming technique, libraries, and corresponding functions.

Two levels of access are established, operator and engineer, who have permission to the existing two processes: assembly and tank control. However, they differ in their accessibility by restrictions for the control of production statistics in the assembly and proportional, integral, and derivative PID control constants of the tank, production graphs, or layout with movement of the assembly process.

The development process and subsequent operation of the human machine interface (HMI) comply with the standardized regulations ISA-SP101 and UNE-EN ISO 9241 according to the characteristics and details shown in Table 2. In summary, strictly necessary, understandable, and clear information is presented, both in tasks and processes, seeking to facilitate the intuitive operation of the process regardless of the user's level of training.

As a result of and in compliance with the standard, the industrial process monitoring website has minimalist features, with a basic development of simulation and realism; therefore, it is a practical HMI interface that allows the identification, deduction, and quick control of the process.

Table 2 Standard ISA-SP101 features

Feature	Description	Detail
Design	Simple	Low contrast colors (balanced)
Order	Logical	Relative alignment of objects
Simulation	Basic	Products and tank level
Text	Clear and simple	Three font sizes, minimal use of capital letters
Controls	Buttons	Grouped by processes (without simulation)
Polarity	Positive	Luminance balance
Colors and shapes	Standard	Background, text, buttons, and information

2.7 *QoS and Security*

The quality of service (QoS) used in the project to define the guarantee of reception of a specific message was mostly "0," due to the characteristics of the network, such as its reliability level and the implemented application logic. In addition, the network proposes a reliable communication and mostly stable connection.

With a QoS level "0," the protocol service guarantees a best-effort delivery; without any delivery guarantee, it provides the same guarantee as the underlying TCP protocol. A QoS level higher than "0" was only used for variables requiring a stable delivery guarantee because of the security and accountability that their functionality represents, such as stop buttons and emergency stop.

In terms of security, the port used in the project is 8883, which allows a higher level of native MQTT security in sending/receiving messages, without any configuration option. Port 8883, standardized specifically for secure MQTT connections, allows encrypted communication of messages both when publishing and subscribing to a specific Topic, using SSL (Transport Layer Security) and TLS (Secure Sockets Layer). SSL and TLS are cryptographic protocols for secure communications, issued by specialized and trusted vendors. The use of these protocols in MQTT causes costs in the use of resources, both economic for the contracting of the service and CPU use and general expenses in the communication network and bandwidth used.

2.8 *Data Traffic Analysis*

Finally, and in parallel to the platform operation tests, its robustness, and response to possible problems, the real functionality of the implemented proposal is determined through the analysis of network traffic of the MQTT protocol using the "Wireshark" tool, which allows us to capture the existing data traffic through Wi-Fi and study the implemented communication.

It also allows the analysis of relevant variables such as bandwidth (bits/s) in the protocol communication, detail of packets in transaction (Paq/s), and packet processing time (Round Trip Time "ms"), in addition to the importance of the analysis and determination of system response times and server load (SBC).

3 Results

The platform has a smooth and dynamic operation, with a perceptive real-time response and admissible characteristics of operability and control. Using the Graph IO tool in Wireshark, an average bandwidth consumption of 225,000 bits per second, i.e., 0.03 MB/s per publish/subscribe communication interaction, is reported. As for

Table 3 Response time of communication

Variable	Description	PLC	Nivel QoS	Response time	
				Wireshark	Cronometer
Bool	Start	%M0.0	0	155 ms	<1 seg
Bool	End emergency	%M3.2	1	267 ms	<1 seg
Int	Set point	%MW28	0	195 ms	<1 seg
Int	Production	%MW32	1	478 ms	<1 seg
Word	Kp	%MD82	0	712 ms	≥1 seg
Word	Td	%MD90	1	932 ms	≥1 seg

Table 4 Raspberry server (broker) test run information

Computer single plate (SBC)				
	SBC Raspberry Pi 3 model B			
	Language:	C	Protocol:	MQTT
	Service:	Broker/Eclipse Mosquitto		
	Communication			
	Client-Raspberry		Variables 60 (28Bool/22Int/10Word)	
	Client-WebSockets			
Functional test				
Date/start:	18/05/2019—10H00	Fate/end:	19/05/2019—16H00	
Operating hours	30 h	Control	Frequent	
Memory RAM	Total: 927 Mb	Used: 146 Mb	Used RAM 17%	
Memory swap	Total: 99 Mb	Used: 0 Mb		
Temperature	Start: 34.9 °C	Stabilization: 35.2 °C		

the analysis of packets and possible delivery/reception errors, an average of 221 packets per second (paq/s) is shown, and the existence of lost data, not useful or with errors, is also verified, a situation that responds to operating on a TCP/IP stack and a QoS level of mostly "0"; however, these packets with errors do not exceed 3% of the total of the publish/subscribe transaction. The average time between packet capture, acknowledgement, filtering, and processing is 600 ms.

In terms of response time, a request/response event presents an average time of 155 ms with the lowest QoS level, a level mostly used in publish/subscribe transactions of the developed platform. This demonstrates the control, communication, and real-time monitoring characteristics of the MQTT protocol. The complementary information of the tests and communication response times is shown in Table 3.

The processing and operating capacity of the client and server hosted on the single-board computers (SBC, Raspberry Pi) now represents the limiting factor to achieve the scalability characteristics of the MQTT protocol. The performance test carried out and detailed in Table 4 allows determining a RAM memory consumption not higher than 20% of its capacity; the swap memory is not used, there is no significant increase in the temperature of the equipment once it is stabilized, and there is no alteration in the normal operation of the process, its control, or monitoring. Consequently, and according to the load applied to the system, its characteristics allow nominally to scale and operate the communication platform through the MQTT protocol, with five clients and around 300 variables of similar characteristics to those established in the tests carried out, taking into account in the latter the controlled and stable environment used.

4 Conclusions

As conclusions of the study, analysis, and performance tests carried out in this chapter, the following can be derived:

The implemented platform based on the concept of Industry 4.0 allows the complete integration of the developed process units, an evident, applicable, and important approach to the industrial transformation so necessary in our country. Evolution is achieved based on the technical knowledge possessed and using open-source devices, adaptable, known, and above all accessible to the user.

MQTT is a relatively easy-to-implement, lightweight, and open messaging protocol that allows communication in restricted environments. It is scalable and bandwidth efficient, enables quality of service levels, and allows message encryption.

The developed platform is online and consists of two clients; the process is controlled by a PLC, simulated in the Factory I/O software, and its communication with the browser client is through WebSockets. After the tests of operation, response, and network traffic analysis, the proposal presents appropriate characteristics of functionality, robustness, scalability, and security.

As a general establishment, the next step to be taken to complement the developed platform is the following:

Perform the application and comparison of MQTT, the current protocol developed in the proposal, with its main competitors in communication, Industry 4.0 and the Internet of Things (IoT), establishing advantages, disadvantages, characteristics, and needs of each protocol, in addition to the specific knowledge required for its proper development and efficient operation.

References

1. C. Ynzunza, J. Izar, J. Bocarando, F. Aguilar, M. Larios, El Entorno de la Industria 4.0: Implicaciones y Perspectivas Futuras. Conciencia Tecnológica **54**, 1–19 (2017)
2. Z. Babovic, V. Milutinovic, *Web Performance Evaluation for Internet of Things Applications*. (IEEEAccess, 2016), pp. 6974–6992

3. J. Val Román, J. Val Román, *Industria 4.0: la transformación digital de la industria* (de CoddInforme, Bilbao, España, 2017)
4. C. Félix, A. Blesa, *Industria 4.0 en el Grado de Ingeniería Electrónica y Automática* (Universitaria de la Informática, 2015), pp. 327–332
5. J. Gonzáles Páramo, Cuarta Revolución Industrial, empleo y Estado de Bienestar, 2017, Real Academia de las Ciencias Morales y Políticas, Madrid, España (2017)
6. R. Oliván, La Cuarta Revolución Industrial, un relato desde el materialismo cultural. Revista de Estudios Urbanos y Ciencias Sociales, 101–111 (2016)
7. J.C. Ortiz, Interviewee, Entrevista, La cuarta revolución industrial. [Entrevista]. (2016)
8. K. Grgic, I. Speh, I. Hedi, *A Web-Based IoT Solution for Monitoring Data Using MQTT Protocol*, in International Conference on Smart Systems and Technologies (SST) (2016), pp. 249–253
9. J. Campero Núñez, ¿La cuarta revolución industrial en Bolivia?, Análisis, pp. 1–24 (2016)
10. Tecnología para los negocios, Sistemas Ciberfísicos: la respuesta a las necesidades de la sociedad y la industria, Diciembre 2018. [En línea]. https://ticnegocios.camaravalencia.com/servicios/tendencias/sistemas-ciberfisicos-la-respuesta-a-las-necesidades-de-la-sociedad-y-la-industria/. [Último acceso: Enero 2019]
11. R. Kodali, K. Mahesh, *A low cost implementation of MQTT using ESP8266*, in 2nd International Conference on Contemporary Computing and Informatics (IC3I) (2016), pp. 404–408
12. B. Lennox, G.A. Montague, H.G. Hiden, G. Kornfeld, P.R. Goulding, Process monitoring of an industrial fed-batch fermentation. Biotechnol. Bioeng. **74**(2), 125–135 (2001). https://doi.org/10.1002/BIT.1102
13. K. Severson, P. Chaiwatanodom, R.D. Braatz, Perspectives on process monitoring of industrial systems. Annu. Rev. Control. **42**, 190–200 (2016). https://doi.org/10.1016/J.ARCONTROL.2016.09.001
14. J. Reyes, L. Morales, D. Aldas, R. Reyes, R. Toasa, Real Time Automatic Andon Alerts for Android Platforms Applied in Footwear Manufacturing, in *Computer and Communication Engineering. ICCCE 2018. Communications in Computer and Information Science*, ed. by M. Botto-Tobar, J. Barzola-Monteses, E. Santos-Baquerizo, M. Espinoza-Andaluz, W. Yánez-Pazmiño, vol. 959, (Springer, Cham, 2019), pp. 43–56. https://doi.org/10.1007/978-3-030-12018-4_4
15. M. León, M. Ruíz, L. Haz, R. Montalvan, V.P. Medrano, S. M. Anchundia, Water Treatment Monitoring System at San Jose de Chaltura, Imbabura—Ecuador, in Lecture *Notes in Computer Science* (including Subser. Lect. Notes Artif. Intell. Lect. Notes Bioinformatics), vol 10963 (LNCS, 2018), pp. 610–624. https://doi.org/10.1007/978-3-319-95171-3_48

Challenges and Benefits for Detecting Soon-to-Fail Drives in Industry 4.0

Marko Krstic, Nicolas Nicolaou, and Efstathios Stavrakis

1 Introduction

Computing and automation in product manufacturing have been key drivers of the third industrial revolution. Building on that foundation, the fourth industrial revolution aims at devising and deploying smarter autonomous systems that are powered by data, analytics, and artificial intelligence.

These cyber-physical systems are no longer considered in isolation; they are seen as part of a distributed, sometimes decentralized, infrastructure for product manufacturing. This design lifts the barriers imposed by physical proximity and enables generation of knowledge through continuous monitoring, data acquisition, and processing from thousands of devices, even in disparate locations. The flow of information from these cyber-physical systems into data analytics and machine learning pipelines is essential for designing adaptive system controls that can respond to the dynamic requirements of production processes, improve operational efficiency, and increase the level of automation in industrial and critical infrastructures [1].

Since data are at the core of such systems, applications in Industry 4.0 rely heavily on reliable, accessible, and dependable data storage. Failures in storage devices of industrial infrastructures may prove catastrophic: computing devices will fail to collect data and perform the necessary computations, thus disrupting the infrastructure control loop. Although the main focus of research in technology ecosystem of Industry 4.0 is usually on communication technologies (e.g., Zigbee, Bluetooth 4.0, SigFox), IoT devices, and computing approaches (e.g., Cloud and Fog computing) [2, 3], it is crucial to develop system for proactive monitoring,

M. Krstic (✉) · N. Nicolaou · E. Stavrakis
Algolysis Ltd, Limassol, Cyprus
e-mail: marko@algolysis.com; nicolas@algolysis.com; stathis@algolysis.com
https://www.algolysis.com

L. Knapčíková, D. Peraković (eds.), *6th EAI International Conference on Management of Manufacturing Systems*, EAI/Springer Innovations in Communication and Computing, https://doi.org/10.1007/978-3-030-96314-9_2

detection, and mitigation of storage device failures—in order to avoid extensive and costly disruptions.

In this chapter, we examine the proactive detection of failing storage devices (or drives), often referred to as *soon-to-fail* drives, by utilizing hardware metrics (i.e., Self-Monitoring, Analysis and Reporting Technology (S.M.A.R.T) measurements [4]) and machine learning (ML) algorithms. S.M.A.R.T. attributes [4] were introduced by storage vendors as a means to examine the condition and usage of drive hardware. These metrics are typically a snapshot of the current status of various attributes of a storage device (e.g., temperature) or cumulative characteristics (e.g., power on hours). Although a snapshot of those metrics may not be sufficient to provide enough insight about the state of a device, a sequence of historical data may reveal patterns that indicate issues on a storage device. Such patterns can be learned by utilizing machine learning models trained with appropriate data.

These metrics can be hard to obtain, and oftentimes due to proprietary technologies used on embedded devices, extracting useful device attributes may not be possible. Furthermore, in the context of Industry 4.0, datasets for training ML models are hardly available. To circumvent those issues and make progress, it is feasible to design models and methodologies for designing such data storage device monitoring and failure prediction systems by drawing from ICT sector. We utilize large-scale datasets of metrics obtained from storage devices in data center infrastructures that have become available (i.e., BackBlaze [5]) and are equivalent in breadth and depth to the data a modern manufacturing facility may generate in the field. We present the methodology of how to preprocess that data and train a number of ML models in TensorFlow framework [6] to proactively detect soon-to-fail drives.

Using the predictive power of these algorithms, we aim to the timely detection of soon-to-fail storage devices in order to: (i) improve the reliability of the existing storage systems by proactively migrating the data to healthy devices before outages and (ii) prevent performance degradation by ill-functioning storage components.

Background The early approaches were trying to predict hard drive failures by setting thresholds for different S.M.A.R.T attributes; however, threshold conservative selection process led to the detection of only 3%–10% of failed drives [7].

The usage of machine learning algorithms for prediction gave significantly better results even when simple techniques are used. For example, supervised naive Bayes classifier trained on a small dataset of 1936 drives (with only 9 failed ones) achieved a prediction accuracy of 33% with 0.67% false alarm rate [8]. Furthermore, in the early phase of the disk failure prediction research, it is shown that more complex algorithms, such as Support Vector Machine (SVM), can provide predictions without false alarms, while detecting 50.6% of failed drives on a small dataset with 369 drives [7]. However, the more representative results emerged with the increased availability of large datasets [5, 9, 10] and interests of large companies (such as Alibaba) that recognized business values of these prediction systems.

With the increased size of datasets and heterogeneity of considered drives, it becomes possible to develop prediction systems that can be used in the production; however, the design of such system is not a trivial task. For example, decision tree model trained on 1 year Backblaze data for all hard drive models available in it can accurately recognize 52% of failed drives, but that comes with a price of 60% false alarms [11]. Sequential modeling of disk failure process by using deep learning models can further improve these results. By using this approach, according to [9], it is possible to achieve not only better prediction performance for a set of disks in Data Centers on which a combination of Convolutional Neural Network (CNN) and Long Short-Term Memory Network (LSTM) is trained, but also the model has much better generalization capabilities on monitoring new Data Centers (not used in training). Furthermore, it is shown that additional information about disk location and system performance (i.e., a number of normal/temp files written successfully) could significantly improve performance as CNN LSTM model trained on Wayne State University dataset can achieve an F1 measure score of 58% when using only S.M.A.R.T. attributes, whereas in case when all available information are used this value goes to 95%. Thus, definition of performance metrics that adequately describes functioning of Industry 4.0 storage systems can be quite beneficial for application of proactive monitoring in these environments. On the other hand, some approaches combine sequential modeling of drive failure process with static information about drives (such as manufacturer and model). The small CNN model with only 6 layers (2 convolutional, 2 maxpooling, and 2 dense) can achieve an F1 measure score of 71% (with false alarms in 25% of cases) if information about drives manufacturer is added into first dense layer and S.M.A.R.T. attributes normalization uses historical ratio of failed and healthy drives for each vendor [12]. However, these information will not always be available and/or behavior of some specific drives used in Industry 4.0 could deviate significantly from regular data center drives; thus using mean values for vendor could introduce significant bias.

In cooperation with academic community, Alibaba as industry partner organized PAKDD (Pacific-Asia Conference on Knowledge Discovery and Data Mining) 2020 AI Ops Competition in order to tackle problem of large-scale disk failure prediction [13]. The general methodology adopted by the most of competitors was comprised of data preprocessing, training sample generation, feature engineering, and modeling [14]. The most commonly used methods for preprocessing were simple ones like dropping the samples with missing data directly, filling them with some constant value (i.e., zero), or interpolating missing data by forward filling methods [15]. The main goal of training sample generation was to reduce class imbalance for what most of participants adopted, either upsampling methods for generation of synthetic samples (i.e., SMOTE [16]) or downsampling methods to randomly select a subset of samples [17] that describe healthy disk functioning. In the feature engineering phase, participants predominately used sliding windows with various window lengths, different statistical measures such as difference, mean, variance, and exponentially weighted moving average values to create features from S.M.A.R.T. attributes, whereas some of them also apply feature selection procedure to remove those features that are not correlated with disk health status. As from the generally used methodology in AI Ops Competition, which includes feature engineering phase, it can be easily concluded, the focus in modeling part was on shallow machine learning algorithms. The disk failure prediction was commonly considered as binary classification problem, and tree-based ensemble models are trained by using either LightGBM [18] or Xgboost [19] framework. The selection of those frameworks is justified by their short execution time as well as low memory requirements that are very important aspects for applications like this when the large amount of data is available.

In addition to mainstream approaches, a few novel methods in each phase of identified general methodology of PAKDD2020 Alibaba AI Ops Competition emerged [14]. For example, it is shown that usage of the cubic spline interpolation method to deal with missing data problem can increase F1 score by more than 3% [20]. The same authors also proposed application of Generative Adversarial Network (GAN) for training sample generation that showed as promising new direction to tackle problem of class imbalance [20]. In the feature engineering phase, a couple of teams proposed feature construction methods based on analysis of the distance of failure occurrences, distributions of disk lifetime, and data missing ratio, which individually but also in the combination led to the improvement of failure prediction [14]. At the end, the possible ways to improve performance in modeling phase are to consider failure prediction as multi-class classification or regression problem [21].

Contributions The main goal of this chapter is to demonstrate the possibilities of utilizing ML algorithms for monitoring and detecting soon-to-fail drives in an Industry 4.0 deployment. To capture a well-rounded and useful outcome, we followed specific steps starting from the identification of challenges all the way to the experimental analysis of the used models. In particular, our contributions are the following:

1. We first identify and present the challenges to overcome for identifying soon-to-fail drives in the wild.
2. Provided the challenges, we then present a methodology for data preparation and propose ML approaches that may be suitable for this problem. More specifically, we consider four different ML models: LSTM and CNN-LSTM that have been used in previous attempts as well, and the ResNET and a variant of a conventional CNN model that are proposed first in this chapter for this application.
3. Due to the nature of the problem and the dataset produced by S.M.A.R.T. metrics, we suggest a methodology for properly training the proposed models.
4. Finally, we implemented and tested our models and obtained experimental results and analytical outcomes for each and every method proposed.

Our results shed some light in the potential of using S.M.A.R.T. metrics alone with powerful ML models to predict any potential drive failures.

Paper Structure This chapter is organized in the following way. Section 2 presents the challenges with which systems for detection of soon-to-fail drives are facing. Sections 3 and 4 describe data preprocessing procedure and prediction models suitable for application in proactive monitoring of storage devices in Industry 4.0, respectively. Training process of ML models is explained in Sect. 5. In Sect. 6, experimental results are presented, whereas finally, Sect. 7 concludes the work.

2 Challenges

Achieving high level of proactive prediction of drive failures is not a trivial problem due to many data-related challenges present in S.M.A.R.T. measurements such as high level of noise, extremely class imbalanced distribution, concept drift phenomena, and a large number and heterogeneity of hard drives included. In more detail, the main challenges to accurate drive failure prediction are the following:

High Level of Noise The noise in datasets for disk failure prediction stems from the non-standardized values of S.M.A.R.T. attributes as well as from the non-reliable and non-appropriate labeling procedure.

Different hard drive vendors may use specific S.M.A.R.T. attributes for different purposes [22]. This introduces noise and makes it much harder for prediction models to learn how healthy drives should function and what are the characteristics that indicate soon-to-fail drives. Such inconsistencies pushed many researchers to use sampling strategies selecting only specific types of disks for training their prediction models and thus avoiding the risk for introducing additional noise through inclusion of many different hard drive models/vendors in training dataset.

On the other side, labeling procedure is equally important both in the context of confirming that reason why hard drive does not send information about S.M.A.R.T. attributes is its failure and that the number of days before failure for which disk is marked as soon-to-fail is chosen in appropriate way, so from S.M.A.R.T. attributes, it is possible to recognize failing conditions. S.M.A.R.T. measurements are also not received in situations when there are some communication problems or if disk itself does not have power supply (either when itself or device using it is turned off, or in case of power outage).

Extremely Class Imbalanced Distributions Due to low hard drive failure rate, the number of samples in which S.M.A.R.T. attributes describe healthy conditions is much bigger than the number of samples that can be used to describe soon-to-fail and failed drives. Therefore, the datasets are characterized by extremely class imbalanced distributions. For example, in the last 3 years, annualized hard drive failure rate in Backblaze dataset (shown in Fig. 1) has values between 1 and 2 % [5]. Furthermore, if we take into account the fact that even for failed disks most of the samples correspond to normal disk functioning conditions and that only during a short period of time before the failure S.M.A.R.T. attributes indicate

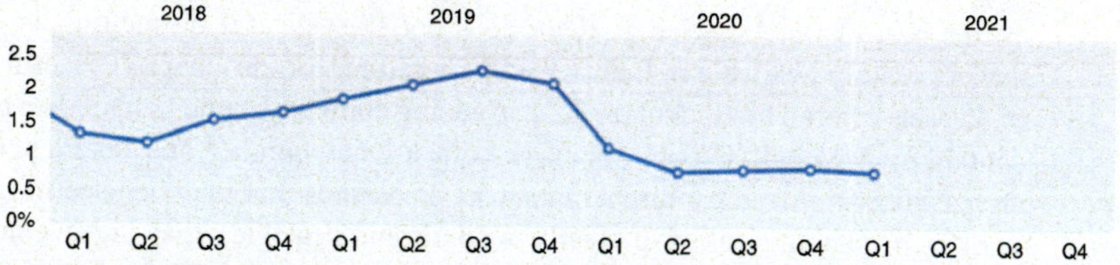

Fig. 1 Annualized hard drive failure rate for Backblaze dataset [5]

failures, it is clearly that without dealing with class imbalance, negative samples will have much bigger impact on the training process. Sampling techniques (such as Synthetic Minority Oversampling Technique—SMOTE [16] that increases the number of positive samples), cost-sensitive learning [23] (that incorporates different weights for positive and negative samples in learning process), special loss functions (with different weights incorporated in them, like focal loss [24]) or anomaly detection approach [25] (that learns how to recognize healthy conditions and detect anomalies) can be used to improve prediction for positive samples (soon-to-fail drives).

Concept Drift Datasets are not only characterized by extremely class imbalanced distributions, but also by concept drift phenomena. In particular, statistical patterns vary in time due to the addition of new and the removal of old or failed drives from the monitored drive pool. To deal with this problem, *change point* detection algorithms can be used to decide when there is a need for retraining or updating the prediction models [26].

The Large Number and Heterogeneity of Hard Drives in a Dataset The problem of hard drive failure prediction can be also considered as Big Data problem due to the large number of hard drives in publicly available datasets, as well as in production systems of big infrastructure providers (such as Alibaba) that monitor large Data Centers. For example, the first quarter of 2021 of the Backblaze dataset contains S.M.A.R.T. measurements for 175,443 drives obtain from four data centers on two continents [5], whereas Wayne State University (WSU) made available a dataset that contains information about 380,000 hard disks over a period of 2 months across 64 sites [9]. Because of this characteristic of datasets, special attention should be paid that prediction models can be trained in reasonable time on computing resources available to the company that develops this type of product. This directly influences which prediction models can be used, the way in which training data is selected, as well as lead to batch-by-batch training process.

In addition to a large number of drives, representative datasets (such as Backblaze and WSU dataset) are also characterized by large heterogeneity. This further introduces more complexity in failure prediction problem as it is not unusual that different drives models have different operating characteristics.

3 Data Processing

To the best of our knowledge, there does not exist dataset on metrics on storage devices readily available in Industry 4.0. Therefore, for this chapter, we utilize a set of datasets offered by Backblaze [22], a leading company in the field of data backup, containing S.M.A.R.T. data obtained from a large number and heterogeneity of drives, allowing us to capture a big subset of devices that may be used in Industry 4.0. Such data may align perfectly with recent strategies that move computing toward the "edge," forming the so-called edge data centers in Smart factories [27].

Backblaze collects S.M.A.R.T. reports from their devices daily. We focused on the data collected in the period between January 1st 2018 and December 31st 2019. This dataset consists of 142138 different drives.

Backblaze reports a drive as failed when it is removed from storage or replaced due to one of the following reasons:

1. The drive has stopped working: This means that it will not power up, does not respond to console commands, or the RAID system alerts that the drive cannot execute read or write operations.
2. The drive is about to fail: Empirical evidence has led to the decision to remove a drive before it fails catastrophically.

In the rest of this chapter, we refer to a *failed* drive when this appears in the period we investigate, and it is marked as failed by Backblaze before the end of the period; otherwise, the drive is *healthy*.

3.1 Dataset Format

Each yearly dataset consists of 365 daily snapshots, which are stored as CSV (Comma Separated Values) files. Each file contains a header row, and a report from each monitored drive for each subsequent row. A drive report represents a daily snapshot of the respective drive. The columns of the dataset represent the following information:

- **Date**: The date of the file in yyyy-mm-dd format.
- **Serial Number**: The manufacturer-assigned serial number of the drive.
- **Model**: The manufacturer-assigned model number of the drive.
- **Capacity**: The drive capacity in bytes.
- **Failure**: Contains a "0" if the drive is healthy. Contains a "1" if this is the last day the drive was operational before failing.
- **Attributes**: 90 columns of S.M.A.R.T. attributes and their normalized values, each associated with an identifier.

3.2 Feature Selection

The sequence of raw S.M.A.R.T. attributes given in Table 1 are used as inputs to our prediction models. The identifiers of S.M.A.R.T. attributes correspond to those used in [9], with the main difference that we use the *raw value* of those attributes that can provide insights in disk functioning without vendor-based normalization. This type of normalization can prevent our ML system to recognize different drive models, what can be really important in the environments with heterogeneous hard drives. Instead, we apply min-max normalization, which transforms the S.M.A.R.T.

Table 1 List of S.M.A.R.T. attributes used

ID	Name	Description
1	Read error rate	Frequency of errors during read operations.
3	Spin-up time	Time required for a spindle to spin up to operational speed.
4	Start/stop count	Raw value holds the actual number of spin-up/spin-down cycles.
5	Reallocated sectors count	The number of the unused spare sectors. When encountering a read/write/check error, a device remaps a bad sector to a "healthy" one taken from a special reserve pool.
9	Power-on hours count	The raw value shows the actual powered-on time, usually in hours.
12	Power cycle count	The raw value holds the actual number of power cycles.
194	Temperature	Temperature, monitored by a sensor somewhere inside the drive. Raw value typically holds the actual temperature (hexadecimal) in its rightmost two digits.

attribute values into the range [0, 1], in order to improve convergence speed of neural networks [28]. For a smart attribute α, the value 0 corresponds to the minimum value of α in our dataset, whereas value 1 corresponds to the maximum value of α.

3.3 Test/Training/Validation Dataset Creation

We generate the testing dataset by selecting 30 healthy and 30 failed drives from each year at random. The remaining data is divided into training and validation datasets in proportion of 70% and 30%, respectively.

With the sequence length parameter, it is possible to control how many previous days S.M.A.R.T. measurements are added to the current to describe operational state of the drive. In this chapter, we adopted a sequence length of 30 days as long enough to model process of disk failing.

Labeling determines whether a drive is about to fail "soon" or whether the drive remains healthy within a predefined time window, termed as *labeling window*. We followed the suggestions presented in [29], where authors analyzed the duration of prefail period for two specific hard drive models from two vendors and determined that 29 and 27 days, respectively, was a satisfying period. Thus, in our experiments, we use a labeling window of 29 days, and we label the records in our dataset in a binary fashion as follows:

- H: A drive record is labeled as *healthy* at a date D, if the drive is not marked as failed in our dataset in any day d s.t., $D < d \leq D + 29$ or we have reached the end of the dataset (whichever is first).
- F: A drive record is labeled as *soon-to-fail* (or failed) at date D, if the drive is marked as failed in the dataset in any day d s.t., $D < d \leq D + 29$.

Based on the label of the drive record, we also classify drives' *operating characteristics*, i.e., the conditions under which a drive operates and which can be measured by the S.M.A.R.T. attribute values. More precisely, if a drive is labeled as *soon-to-fail* (F), then the values of S.M.A.R.T. attributes describe soon-to-fail operating characteristics; otherwise, the S.M.A.R.T. values describe healthy drive operating characteristics.

In order to deal with the class imbalance problem, healthy drives are randomly selected from our dataset in such a way that the ratio between failed and healthy ones, in both training and validation sets, was 1:2. This ratio is selected in order to reduce class imbalance but also to preserve diversity of drives for training the ML models. However, as each drive can be operating for different time durations, resulting into a different number of samples per drive, additional *undersampling and cost-sensitive learning methods* will be examined to further reduce class imbalance (see Sect. 5.2).

4 Prediction Models

In this section, we present the ML models we consider, and we substantiate their applicability in Industry 4.0 applications.

4.1 LSTM and CNN LSTM Models

Industry 4.0 environments often include rare and non-typical computing devices that may use non-conventional hard drives as storage mediums. To this end, in order to provide viable proactive solutions for detecting soon-to-fail drives in such environments, we should examine solutions that model drive failures through a sequence of historical operating characteristics, an approach that showed good generalization capabilities especially in the case of LSTM and CNN LSTM models [9]. LSTM model learns directly from S.M.A.R.T. attributes whose patterns correspond to soon-to-fail conditions. On the other hand, CNN LSTM model uses CNN for features extraction and LSTM to support sequence classification. An illustration of the structure of LSTM and CNN LSTM models is shown in Fig. 2.

LSTM layers [31] are composed of LSTM units that can process data sequentially and save hidden state ($h(t)$) through time. The functioning of LSTM unit is described in Fig. 2. With symbols $x(t)$ the input vector at timestep t, $h(t-1)$ the hidden state value in timestep $(t-1)$, $c(t)$ and $c(t-1)$ the cell state vectors in the current timestep t and in the previous timestep $(t-1)$, and o the Hadamard product. Initial values for cell state vector $c(0)$ and hidden state $h(0)$ are zeros, whereas the other timesteps in sequence are computed according to Fig. 2. LSTM unit contains input, forget, and output gates that control its behavior, and makes the propagation of unchanged gradients from previous timesteps possible, in order to support deeper

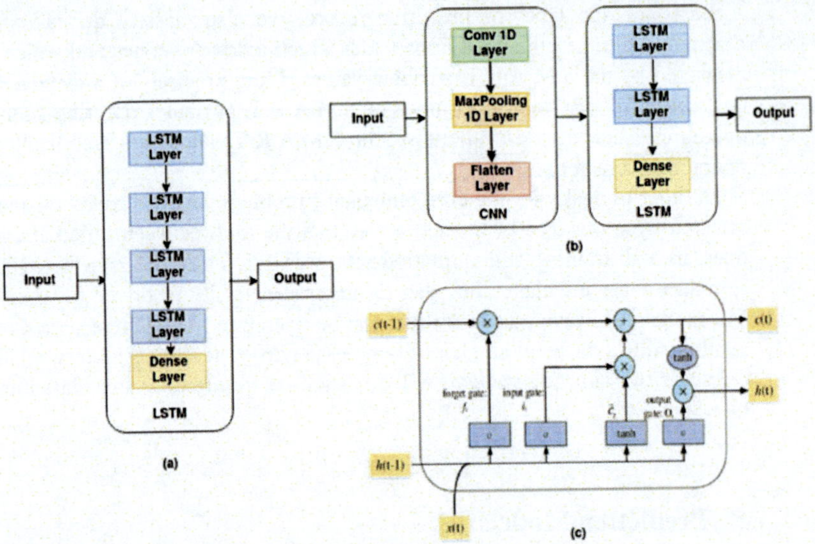

Fig. 2 (**a**) LSTM model [9], (**b**) CNN LTSM model [9], and (**c**) LSTM Unit [30]

neural network architectures. On the top of the LSTM layers in the LSTM model [9], fully connected Dense layer [32] is added in order to support binary classification function (healthy or soon-to-fail drives), what is the main goal of the neural network as a whole.

On the other hand, CNN LSTM model [9] applies additional transformation before LSTM layers. Conv 1D layer [33] first calculates convolutions over time dimension, then MaxPooling 1D [34] layer downsamples the input representation by taking the maximum value over spatial windows, whereas last Flatten layer [35] transforms its input vector into the shape (batch size, sequence length, the number of features) that is expected by LSTM layers.

4.2 The ResNet Model

As in its essence modeling failures by a sequence of historical operational characteristics translates disk failure prediction problem into the so-called Time Series Classification (TSC) problem [36], proactive failure detection systems could benefit from the recent advancement made in TSC area.

Until recently, hierarchical VotE—Collective Of Transformation-based Ensembles (HIVE-COTE) [37] that combines predictions from 35 individual classifiers built on four different data representations—was recognized as the only way to achieve state-of-the-art performance in TSC problems; however, due to its complexity, it cannot be trained in reasonable time for large datasets. In these

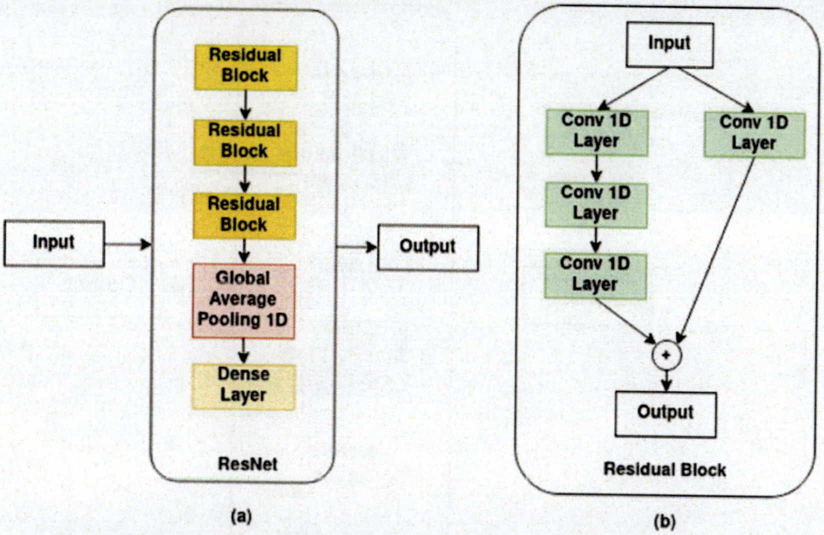

Fig. 3 **(a)** Structure of ResNet model [38]. **(b)** Structure of residual block [1]

scenarios, CNNs emerged as great alternative because of their capability to learn time-invariant representations [38].

The ResNet architecture [36], with 3 residual blocks followed by a global average pooling layer [39] (that averages the time series across the time dimension), showed its clear dominance in wide spread of domains for both univariate and multivariate time series [38]. The structures of ResNet model and residual block are shown in Fig. 3.

The time-invariant representation of S.M.A.R.T. attributes that ResNet (and generally CNNs) generates enables failure prediction systems to detect important patterns that are not time dependent—something really important as different hard drives can exhibit similar patterns but not at the same time interval before the failure.

4.3 The Simple CNN Model

The computational resources (i.e. processing power and memory) required in order to deploy a prediction model on low-powered computing devices are an important factor to consider in the context of Industry 4.0. To further explore this aspect, we modified the simple CNN architecture [12] that combines sequential (S.M.A.R.T. attributes) and static information (disk vendor), by removing the dense layer used to include vendor information. Information about the manufacturer could help the ML model to differentiate between the behavior of drives from different vendors. However, in order to be successful, each vendor must be represented

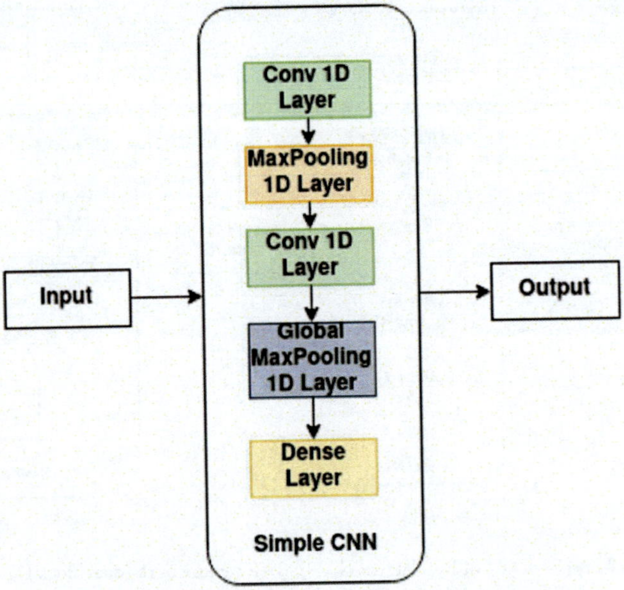

Fig. 4 Structure of Simple CNN model

by a "sufficient" number of measurements in the dataset. As this prerequisite is not always fulfilled in Industry 4.0, we modified the ML model architecture. The structure of this model, named *Simple CNN* hereinafter, is shown in Fig. 4. The main difference between Global MaxPooling 1D [40] and MaxPooling 1D Layer [34] in Simple CNN architecture is that when downsample input representation first method finds the maximum value in the whole time axis—thus reducing the dimensionality of the layer output vector by one, the latter considers maximum values in different windows among time axis. This reduction in vector dimensionality is needed in order to prepare the inputs for the Dense layer.

5 Training Process

Adam, an adaptive learning rate optimization algorithm, with default initial learning rate parameter (lr=0.001) [41] was used in the process of training all prediction models described in Sect. 4.

Early stopping criteria, which will stop training if the optimization loss of the model on the validation dataset increases in the predefined number of consecutive complete passes of training data (the so-called training epochs), are applied to avoid model overfitting. The training is done iteratively—so the training data is presented to neural network multiple times. We say that one training epoch is done when all the samples in a training dataset are presented to the neural network. The training dataset is divided into chunks of data (the so-called batches [42])—when one batch is presented to the NN—we say that one iteration of training is done. After each epoch, model performance is evaluated on the whole validation dataset. If the loss increases or the value of some metrics (e.g., accuracy) declines for consecutive epochs, this means that further training will negatively affect the performance of the model on unseen datasets (although the performance on the training dataset could be improved). So in such a case, we stop the training. Have we not apply this method, we increase the possibility of encountering the risk of achieving good performance for the training dataset and poor generalization capability on unknown datasets.

Having in mind the challenges that proactive monitoring systems of storage devices are facing, special attention was paid to methods that deal with the large amounts of data and with class imbalance.

5.1 *Dealing with the Large Amounts of Data*

As the size of the representative dataset is usually larger than the size of the RAM in a computing device (especially for low spec devices as those used in Industry 4.0), the training of ML models is typically done by splitting data into batches and iteratively processing a single batch at a time. None of the previous works in the literature on proactive hard drive monitoring provided guidelines on how training batches should be constructed. Thus, in this chapter, we examined two approaches where each batch is formed by:

1. Data collected for a single drive
2. A combination of data from failed and healthy drives

In both cases, *sampling without replacement* is applied to select the drives of which the S.M.A.R.T. measurements will be considered in the batches. This ensures that each drive from the population has only one chance of being included in the batch. The main difference between (1) and (2) is that in the first case, for each batch only one drive is selected, whereas in the latter case first a failed drive is chosen, followed by two healthy drives, and then their data are combined. The ratio of failed and healthy drives in latter case is used as it corresponds to the ratio initially applied to the downsampled dataset.

5.2 *Dealing with Class Imbalance*

Once we split our dataset into batches, we then need to address the issue of class imbalance.

In the case where a batch is formed from data collected for a single drive, we apply both undersampling and cost-sensitive learning methods. For every healthy disk in the training dataset, 64 data points are randomly selected, whereas for failed ones, all data points that correspond to soon-to-fail operating characteristics are included with addition of 34 randomly selected data points from times when they did function properly. The class imbalance factor is then calculated as the ratio between the number of data samples where drives are marked as healthy and those marked as soon-to-fail, and in the process of learning impact of errors on healthy drives (negative) samples is reduced by this factor.

On the other hand, in the case when batch is formed from both healthy and failed drives data, we did not use undersampling to further reduce class imbalance in the training dataset. The reason for this decision was to support learning of differences between healthy and soon-to-fail operating characteristics in the same batch with as many examples. However, in addition to standard cost-sensitive learning method, described in paragraph above, the advanced method to fight with class imbalance in ML models with sequential inputs is also considered. Recent research showed that if cost/weights in training process are adjusted dynamically to class imbalance ratio in each batch, instead of global class imbalance ratio, ML models can achieve better classification performance [43], but until now this method is not examined in the context of the application in the disk failure prediction.

6 Experimental Results

To evaluate prediction capabilities and select the best neural network architecture for the proactive detection system for soon-to-fail drives, we conducted a series of experiments. As performance metrics, *accuracy*, *precision*, *recall*, and *F1 score* were selected. Those are defined by the following equations:

$$accuracy = (TP + TN)/(TP + TN + FP + FN) \tag{1}$$

$$precision = TP/(TP + FP) \tag{2}$$

$$recall = TP/(TP + FN) \tag{3}$$

$$F1 = 2 * (recall * precision)/(recall + precision), \tag{4}$$

where TP, TN, FP, and FN are the number of true positives, true negatives, false positives, and false negatives, respectively.

The prediction model weights that achieve the best F1 measure for validation dataset were saved as final output of the training for each model, and those are used for model performance comparison.

The experiments are divided into those that explore:

1. Different ways to generate data in training batches
2. Advanced class imbalance technique for time series.

Table 2 Performance of models in case when batches are formed from one disk data

Model	Precision	Recall	F1 measure	Accuracy
LSTM	0.45	1	0.62	0.55
CNN LSTM	0.45	1	0.62	0.55
ResNet	0.57	0.23	0.32	0.57
Simple CNN	0.56	0.8	0.66	0.6

Table 3 Performance of models in case when batches are formed from a mixture of failed and non-failed drives

Model	Precision	Recall	F1 measure	Accuracy
LSTM	0.57	0.88	0.69	0.61
CNN LSTM	0.57	0.88	0.69	0.61
ResNet	0.36	0.6	0.45	0.55
Simple CNN	0.32	0.66	0.43	0.52

6.1 Different Ways to Generate Data in Training Batches

In the first series of experiments, the performances of LSTM, CNN LSTM, ResNet, and Simple CNN neural network architectures are compared for cases when each batch is formed from one disk data and when combine data from failed and healthy disks, respectively. The corresponding results are given in Tables 2 and 3.

According to Table 2, in the case when each batch is formed from one drive data, the best F1 measure can be achieved by using Simple CNN architecture. This model is capable of finding 80% of all failed drives; however, this comes at price of 44% of false alarms. On the other side, the more complex architectures like LSTM and CNN LSTM are able to recognize all soon-to-fail drives in test dataset, but their precision is significantly reduced in comparison to Simple CNN architecture. There was no difference in the performances of those models, so we can conclude that in our case representations learned by CNN were not more informative than S.M.A.R.T. attributes as inputs to LSTM network. Lastly, the most complex architecture Resnet performs poorly in the case when data from just one drive is included in each batch—even precision is slightly better than for Simple CNN architecture, it is capable to detect only 23% of failed drives in test dataset.

If we compare results from Tables 2 and 3, we can easily observe that in the case when batches are formed as a mixture of failed and non-failed drives, the performance of more complex models is improving, whereas getting worse for Simple CNN model. The best performance under this condition is achieved by LSTM and CNN LSTM models. Those are capable to detect 88% of drives that are in soon-to-fail conditions, while generating false alarms in 43% of cases. Although the F1 measure achieved by ResNet architecture was significantly increased, this model is still not one of the best ones. By using batches that contain information about both soon-to-fail and healthy disks, it is possible to significantly increase recall of ResNet model, but at the same time precision is deteriorated. On the other hand, this way of forming batches apparently increases complexity of learning due to the mixture of failed and non-failed drives in each batch. Although that can

Table 4 Performance of models in case when different class imbalance weights/costs are incorporated at batch level

Model	Precision	Recall	F1 measure	Accuracy
LSTM	0.51	0.79	0.61	0.55
CNN LSTM	0.51	0.79	0.61	0.55
ResNet	0.34	0.54	0.37	0.51
Simple CNN	0.3	0.58	0.4	0.53

be beneficial for models with a deeper architecture, it may lead to performance degradation (in both precision and recall) for simple models like the Simple CNN.

6.2 Advanced Class Imbalance Technique for Time Series

In the second series of experiments, impact of introducing dynamic weights [43] into learning process to deal with class imbalance in the case when batch is formed as a mixture of failed and non-failed drives is examined. The performances that different neural network architectures achieved under this condition are given in Table 4.

Unlike expected, dynamic adjustment of weights of cost-sensitive learning to distribution in each batch of mixed failed and non-failed drives did not led to performance improvement, as it can be observed from Table 4. Even more, the performances for all neural network architectures were degraded. The possible reason for such behavior can be direct relation between global class imbalance in dataset and annual failure rates of drives, whereas this relation is not so clear in case when the weight/cost for positive and negative samples is adjusted in batch-by-batch basis. Thus global weights/cost besides helping models to deal with class imbalance could possibly incorporate information about global annual failure rates of drives used in training dataset.

6.3 Discussion

Bearing in mind, results of all experiments described above, LSTM model (shown in Fig. 2) should be trained in batch-by-batch manner with batches formed as a mixture of healthy and failed drives in order to achieve best performance of system for detection of soon-to-fail drives. Although CNN LSTM model could achieve the same performance, it introduces additional complexity that can be issued in some implementations of Industry 4.0. The model complexity can be really important in deployments where edge computing paradigm is used, and ML models are implemented locally on processing devices in Smart Factories. For such cases, Simple CNN model trained with batches of data from one-by-one drive can be an appropriate alternative.

7 Conclusion

As Industry 4.0 applications are heavily dependent on data acquisition and processing, storage devices in industrial infrastructure should fulfill strict requirements in terms of reliability, performance, and monitoring. However, regardless of this fact, the applications of proactive monitoring and detection systems, already considered by big infrastructure providers like Alibaba, are not yet thoroughly explored in Industry 4.0.

In this chapter, we demonstrated applicability of proactive detection systems for soon-to-fail drives in Industry 4.0 environments. The hard drive failure process is modeled as a sequence of drive operational characteristics measured in the last 30 days by using ML models. Although specialized datasets for hard drives in Industry 4.0 still do not exist, the large number and heterogeneity of drives in Backblaze dataset was adequate to emulate those environments in which similar characteristics are expected. The proposed detection system is designed in such a way to overcome challenges that arise from data characteristics, with the main focus on Big Data and class imbalance implication. It is experimentally shown that LSTM neural network, trained in batch-by-batch manner, where each batch is formed from both information about healthy and failed drives, is capable to recognize 88% of drives that are in soon-to-fail conditions. However, this comes at a price of 43% of false alarms; thus further research is needed to improve prediction precision. In order to achieve this, future research will consider combining different ML models and ways of data representation (a sequence of measurements vs. only last measurement). The recent advances from Time Series Classification area in terms of ML models and methods to deal with class imbalance, which we examined in this chapter, did not show valuable for application in domain of hard drive failure prediction.

Acknowledgments This work was partially supported by the EU H2020 Innovation Associate grant PREFAIL (957149) and by the Cyprus Research and Innovation Foundation under the grant agreement POST-DOC/0916/0090.

References

1. Y. Lu, Industry 4.0: A survey on technologies, applications and open research issues. J. Ind. Inf. Integr. **6**, 1–10 (2017)
2. D. Peraković, M. Periša, R.E. Sente, Information and communication technologies within Industry 4.0 concept, in *Design, Simulation, Manufacturing: The Innovation Exchange* (Springer, Berlin, 2018), pp. 127–134
3. D. Peraković, M. Periša, P. Zorić, Challenges and issues of ICT in Industry 4.0, in *Design, simulation, manufacturing: The innovation exchange* (Springer, Berlin, 2019), pp. 259–269
4. S.S. KM, et al., Self monitoring analysis and reporting technology (SMART) copyback, in *International Conference on Information Processing* (Springer, Berlin, 2011), pp. 463–469
5. Backblaze dataset. https://www.backblaze.com/b2/hard-drive-test-data.html, last accessed 2021/07/13
6. TensorFlow framework. https://www.tensorflow.org/, last accessed 2021/08/01

7. J.F. Murray, G.F. Hughes, K. Kreutz-Delgado, D. Schuurmans, Machine learning methods for predicting failures in hard drives: A multiple-instance application. J. Mach. Learn. Res. **6**(5), 783–816 (2005)
8. G. Hamerly, C. Elkan, et al., Bayesian approaches to failure prediction for disk drives, in *ICML*, vol. 1 (Citeseer, New York, 2001), pp. 202–209
9. S. Lu, B. Luo, T. Patel, Y. Yao, D. Tiwari, W. Shi, Making disk failure predictions SMARTer!, in *18th USENIX Conference on File and Storage Technologies (FAST 20)* (2020), pp. 151–167
10. Alibaba PAKDD2020 dataset. https://github.com/alibaba-edu/dcbrain/tree/master/diskdata, last accessed 2021/07/13
11. C.A. Rincón, J.F. Pâris, R. Vilalta, A.M. Cheng, D.D. Long, Disk failure prediction in heterogeneous environments, in *Proceedings of the 2017 International Symposium on Performance Evaluation of Computer and Telecommunication Systems (SPECTS)* (IEEE, New York, 2017), pp. 1–7
12. X. Sun, K. Chakrabarty, R. Huang, Y. Chen, B. Zhao, H. Cao, Y. Han, X. Liang, L. Jiang, System-level hardware failure prediction using deep learning, in *Proceedings of the 2019 56th ACM/IEEE Design Automation Conference (DAC)* (IEEE, New York, 2019), pp. 1–6
13. PAKDD2020 Alibaba AI Ops Competition. https://tianchi.aliyun.com/competition/entrance/231775/introduction?lang=en-us, last accessed 2021/07/13
14. C. He, Y. Liu, T. Huang, F. Xu, J. Liu, S. Han, P.P. Lee, P. Wang, Summary of PAKDD CUP 2020: From organizers' perspective, in *AI Ops Competition* (Springer, Berlin, 2020), pp. 130–142
15. R.N. Faizin, M. Riasetiawan, A. Ashari, A review of missing sensor data imputation methods, in *Proceedings of the 2019 5th International Conference on Science and Technology (ICST)*, vol. 1 (IEEE, New York, 2019), pp. 1–6
16. N.V. Chawla, K.W. Bowyer, L.O. Hall, W.P. Kegelmeyer, SMOTE: synthetic minority over-sampling technique. J. Artif. Intell. Res. **16**, 321–357 (2002)
17. X. Ran, Z. Su, Anomaly detection of hard disk drives based on multi-scale feature, in *AI Ops Competition* (Springer, Berlin, 2020), pp. 40–50
18. G. Ke, Q. Meng, T. Finley, T. Wang, W. Chen, W. Ma, Q. Ye, T.Y. Liu, LightGBM: A highly efficient gradient boosting decision tree. Adv. Neural Inf. Proces. Syst. **30**, 3146–3154 (2017)
19. T. Chen, C. Guestrin, XGBoost: A scalable tree boosting system, in *Proceedings of the 22nd ACM SIGKDD International Conference on Knowledge Discovery and Data Mining* (2016), pp. 785–794
20. Q. Wu, W. Chen, W. Bao, J. Li, P. Pan, Q. Peng, P. Jiao, Tree-based model with advanced data preprocessing for large scale hard disk failure prediction, in *AI Ops Competition* (Springer, Berlin, 2020), pp. 85–99
21. J. Zhang, Z. Sun, J. Lu, First place solution of PAKDD CUP 2020, in *AI Ops Competition* (Springer, Berlin, 2020), pp. 30–39
22. N. Aussel, S. Jaulin, G. Gandon, Y. Petetin, E. Fazli, S. Chabridon, Predictive models of hard drive failures based on operational data, in *Proceedings of the 2017 16th IEEE International Conference on Machine Learning and Applications (ICMLA)* (IEEE, New York, 2017), pp. 619–625
23. N. Thai-Nghe, Z. Gantner, L. Schmidt-Thieme, Cost-sensitive learning methods for imbalanced data, in *The 2010 International Joint Conference on Neural Networks (IJCNN)* (IEEE, New York, 2010), pp. 1–8
24. T.Y. Lin, P. Goyal, R. Girshick, K. He, P. Dollár, Focal loss for dense object detection, in *Proceedings of the IEEE International Conference on Computer Vision* (2017), pp. 2980–2988
25. R. Chalapathy, S. Chawla, Deep learning for anomaly detection: A survey. arXiv preprint arXiv:1901.03407 (2019)
26. S. Han, P.P. Lee, Z. Shen, C. He, Y. Liu, T. Huang, Toward adaptive disk failure prediction via stream mining, in *Proceedings of the 2020 IEEE 40th International Conference on Distributed Computing Systems (ICDCS)* (IEEE, New York, 2020), pp. 628–638
27. Top tips for IIOT storage in Industry 4.0. https://technative.io/top-tips-for-iiot-storage-in-industry-4-0/, last accessed 2021/08/01

28. V.S. Durga, T. Jeyaprakash, An effective data normalization strategy for academic datasets using log values, in *Proceedings of the 2019 International Conference on Communication and Electronics Systems (ICCES)* (IEEE, New York, 2019), pp. 610–612
29. S. Han, J. Wu, E. Xu, C. He, P.P. Lee, Y. Qiang, Q. Zheng, T. Huang, Z. Huang, R. Li, Robust data preprocessing for machine-learning-based disk failure prediction in cloud production environments. arXiv preprint arXiv:1912.09722 (2019)
30. X. Yuan, L. Li, Y. Wang, Nonlinear dynamic soft sensor modeling with supervised long short-term memory network. IEEE Trans. Ind. Inf. **16**(5), 3168–3176 (2019)
31. LSTM layer. https://www.tensorflow.org/api_docs/python/tf/keras/layers/LSTM, last accessed 2021/08/01
32. Dense layer. https://www.tensorflow.org/api_docs/python/tf/keras/layers/Dense, last accessed 2021/08/01
33. Conv1D layer. https://www.tensorflow.org/api_docs/python/tf/keras/layers/Conv1D, last accessed 2021/08/01
34. Maxpooling 1D layer. https://www.tensorflow.org/api_docs/python/tf/keras/layers/MaxPool1D, last accessed 2021/08/01
35. Flatten layer. https://www.tensorflow.org/api_docs/python/tf/keras/layers/Flatten, last accessed 2021/08/01
36. Z. Wang, W. Yan, T. Oates, Time series classification from scratch with deep neural networks: A strong baseline, in *Proceedings of the 2017 International Joint Conference on Neural Networks (IJCNN)* (IEEE, New York, 2017), pp. 1578–1585
37. J. Lines, S. Taylor, A. Bagnall, HIVE-COTE: The hierarchical vote collective of transformation-based ensembles for time series classification, in *Proceedings of the 2016 IEEE 16th International Conference on Data Mining (ICDM)* (IEEE, New York, 2016), pp. 1041–1046
38. H.I. Fawaz, G. Forestier, J. Weber, L. Idoumghar, P.A. Muller, Deep learning for time series classification: a review. Data Min. Knowl. Disc. **33**(4), 917–963 (2019)
39. Global average pooling 1D layer. https://www.tensorflow.org/api_docs/python/tf/keras/layers/GlobalAveragePooling1D, last accessed 2021/08/01
40. Global max pooling 1D layer. https://www.tensorflow.org/api_docs/python/tf/keras/layers/GlobalMaxPool1D, last accessed 2021/08/01
41. Adam optimizer. https://www.tensorflow.org/api_docs/python/tf/keras/optimizers/Adam, last accessed 2021/08/01
42. B. Liu, W. Shen, P. Li, X. Zhu, Accelerate mini-batch machine learning training with dynamic batch size fitting, in *Proceedings of the 2019 International Joint Conference on Neural Networks (IJCNN)* (IEEE, New York, 2019), pp. 1–8
43. Y. Geng, X. Luo, Cost-sensitive convolution based neural networks for imbalanced time-series classification. arXiv preprint arXiv:1801.04396 (2018)

End Users and Industry 4.0 Systems Cyber Resilience from XSS Attacks

Ivan Cvitić, Dragan Peraković, Siniša Husnjak, and Dominik Sever

1 Introduction

In the contemporary world, it is hard to imagine everyday life without access to the Internet. It is an unavoidable part of both free time and work. Moreover, an increasing number of companies provide their services over the Internet. All of these make everyday work easier for the users [1]. The problem arises when there is an increased risk of cyberattacks with the increasing interaction over the Internet. Any user connected to the Internet can be the target of a cyberattack, whether it is a private user who uses the Internet in their free time or business users and companies.

The modern concept of Industry 4.0 is characterized by the interconnection of various elements, business processes, and usage of information and communication technologies, concepts such as the Internet of Things, cloud computing, and big data [2]. In such an environment, many cyber threats can compromise the industrial management system, cause interruption or stop the business process, and jeopardize human safety or cause considerable financial losses [3]. Some of the most commonly used attacks are distributed denial of service (DDoS), social engineering (including phishing attacks), injection, and cross-site scripting (XSS) attacks [4, 5].

One of the most common types of cyberattacks is XSS attacks. With XSS, the attacker can take advantage of seemingly harmless web pages to implement the malicious script, making them hard to detect for the end user [6]. Therefore, it is necessary to frequently work on researching this type of cyberattack to analyze and improve existing security solutions, develop new solutions, and generally point out the dangers behind XSS attacks. This research aims to simulate XSS attack in

I. Cvitić (✉) · D. Peraković · S. Husnjak · D. Sever
Faculty of Traffic and Transport Sciences, University of Zagreb, Zagreb, Croatia
e-mail: ivan.cvitic@fpz.unizg.hr; dragan.perakovic@fpz.unizg.hr; sinisa.husnjak@fpz.unizg.hr

L. Knapčíková, D. Peraković (eds.), *6th EAI International Conference on Management of Manufacturing Systems*, EAI/Springer Innovations in Communication and Computing, https://doi.org/10.1007/978-3-030-96314-9_3

several scenarios using the BeEF framework, and through synthesized simulation results proposes guidelines for improving end-user resilience on such attacks.

The rest of the chapter is organized as follows. In the second section, related research is analyzed to position our research and contribution. In the third section, XSS attack is analyzed, its work principles are described, and classification of XSS attacks and trends is provided. In the fourth section, the XSS attack is simulated using the BeEF framework in several scenarios. In the fifth section, simulation results are synthesized, and guidelines for end-user protection from XSS attacks are provided. In the final section, the research conclusion is given.

2 Related Research

As said before, it is important to work on solutions that can secure users from XSS attacks. In many studies on XSS attacks, various solutions and methods for defense against attacks have been proposed. In research [7], the authors suggested defense method by applying filtering based on defined samples. It is suggested to apply filtering on event handlers, uniform resource identifiers (URI), unsafe keywords, and special characters. The authors gave an example of filtering where users' input is analyzed for predefined list of keywords (for example: *document.cookie, document.write*). If the user's input contains some of the keywords, it is most likely an XSS attack so the input will not be stored in the database [7].

A similar way of filtering is applied to other previously mentioned elements. The authors of research [8] proposed a method called XSSDS (server-side detection of cross-site scripting). The proposal for defense against reflected XSS attacks relies upon the observation that there is a strong correlation between input parameters and reflected XSS issues. In other words, if an attacker injects a malicious script in HTTP request, that script should be fully contained in HTTP response. Therefore, detecting of XSS can be done by matching input and output data and checking for similarity. This approach cannot be used for stored XSS, so for that type of XSS attack authors suggest using a detector that will analyze all outgoing JavaScript scripts with the list of known scripts. If there is a variation detected, some new, unknown script is most likely malicious [8].

A similar defense method is suggested in research [9]. The authors suggest a solution called Secure Web Application Proxy (SWAP). It is a server-side solution that works as a reverse proxy because every data sent from the server back to the user must pass through that reverse proxy. In it, there is an implemented detector. If a detector detects an unknown script sent to the user, the response is blocked [9].

From the analyzed research, the potential threat of XSS attacks on Industry 4.0 and similar environments is visible, and a strong need for further research and simulations of such attacks to provide guidelines for protecting both end users and the system as a whole is highlighted.

3 Analysis and Classification of XSS Attacks

Cross-site scripting attacks are a type of cyberattack that is performed when an attacker implements a malicious code or script into a seemingly harmless website or web application. Malicious scripts are usually written in the JavaScript programming language but can also be written in other languages. JavaScript is a language that is executed in the user's web browser and allows dynamic elements of the page. Therefore, the attack itself occurs not only by implementing a malicious script in the page but also by executing it in the victim's browser when visiting the specified website or application [10].

The target of the XSS attack is the end user of the website, not the server, as is the case with some cyberattacks. Therefore, it can be said that a website with malicious code is only an intermediary that allows the implementation of the code and its transfer to the victim. Once an attacker takes control of a user's web browser, he/she can perform many dangerous actions such as stealing user data, keylogging, browsing search history, stealing cookies, and accessing many other sensitive data that can be used in further illegal actions [11].

3.1 Classification of XSS Attacks

XSS attacks can be classified into three categories:

- Stored or persistent XSS
- Reflected or nonpersistent XSS
- DOM-based XSS

In case of a stored XSS attack, an attacker implements his/her malicious script into the website and the script is then stored into the database. Stored XSS attacks occur less frequently than reflected ones. Today, it is relatively difficult to find a website or application that is vulnerable to this type of attack. As malicious code is stored on the server, any visitor to the site can be the victim of an attack, not just a specific person who is sent a link to the page [12]. Commonly most vulnerable types of websites for this kind of XSS are forums, blogs, social networks, and basic websites with some field for user input, like comments [13].

Reflected XSS occurs when the malicious script is not stored on the web server but is reflected back to the victim as a response from a web server. The attacker crafts a URL (Uniform Resource Locator) and sends it to a victim. For example, a URL can contain a search on a specific website but with a malicious script as an argument. When an attacked user clicks on the link, he/she receives a response from a web server containing a malicious script, which is then executed in his/her browser [12].

The third category is DOM-based XSS. This type of XSS works by executing the attacker's payload due to changing the DOM environment in the victim's browser. DOM-based XSS can seem very similar to reflected XSS because it can also be conducted by sending a malicious URL to the attacked user, but one of the differences is that in case of DOM-based XSS, the response from the server does not contain an attacker's payload, and it manifests at a client side [14].

3.2 Trends in XSS Attacks

Trends and statistics on cyberattacks are an important indicator of the severity of the problems posed by certain types of attacks and provide a basis for further research. Many organizations and companies dealing with cyber security collect data from their research and publish them in the report. For example, the company WhiteHat Security states that XSS was the second most common vulnerability in web applications in 2017, with the likelihood of 33%. That data was gathered with dynamic analysis [15]. Another example of frequency of XSS attacks can be seen from the data in Figs. 1 and 2. Figure 1 shows most common cyberattacks in 2017, and Fig. 2 shows most common cyberattacks in 2018. The data is taken from [16, 17]. As it can be seen, in both years XSS attacks are in top three most common attacks alongside SQL injection and path traversal. Of course, there are variations between these three types, but they are always at the top of the list.

Newer data is shown in Fig. 3. Data in Fig. 3 is taken from [18] and shows the most common critical risk vulnerabilities discovered by Edgescan in 2020.

As it can be seen from Fig. 3, XSS vulnerabilities are again in the top three of the list. If figures shown above are analyzed, it can be noticed that XSS attacks are at the very top both as a form of vulnerability and in the number of attacks performed. Therefore, it can be concluded that XSS attacks are a widespread form of cyberattacks chosen by attackers and their frequency does not decrease significantly over the years.

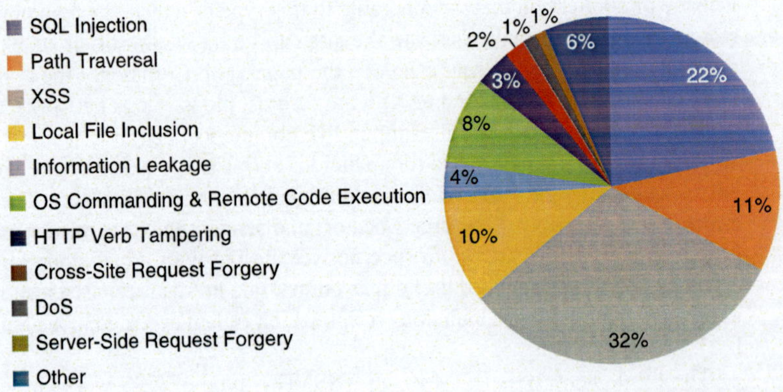

Fig. 1 Most common cyberattacks in 2017 [21]

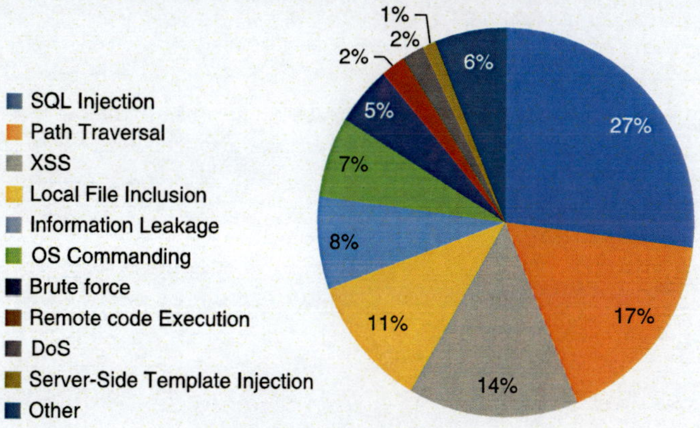

Fig. 2 Most common cyberattacks in 2018

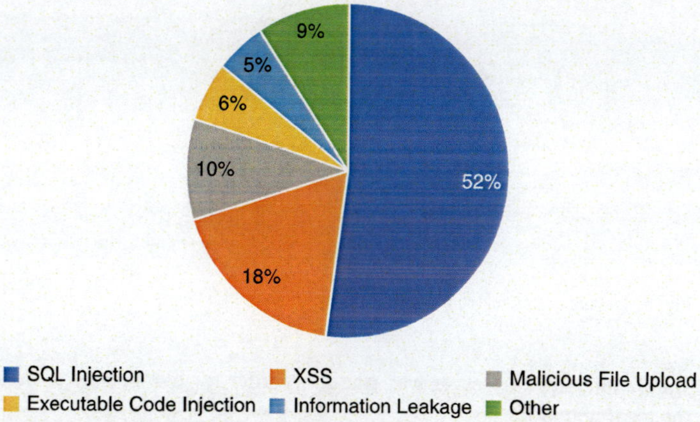

Fig. 3 Most common critical risk vulnerabilities in 2020

4 XSS Attack Simulation Using a BeEF Framework

For this research, a simulation of XSS attack was performed. The main tool used for the simulation was BeEF framework. It is a tool intended for security testing, which, as its name suggests, is based on the use of "open doors" in every computer, and that is a web browser. Once the web browser is compromised, various attacks can be performed through the tool interface, depending on the level of protection on the attacked side [19]. Figure 4 shows a simple diagram of the way BeEF is used.

In this simulation the "attacker" is a laptop with BeEF installed and the "attacked browser" is a browser installed on the second PC on a local network. The Hook.js provided by BeEF is implemented into a simple HTML website created for this

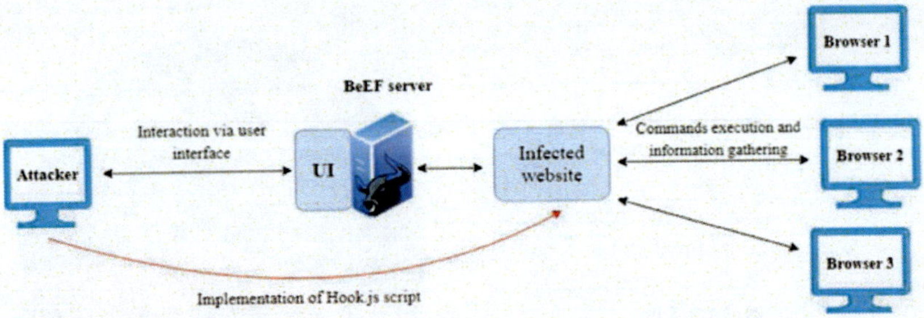

Fig. 4 High-level view of BeEF

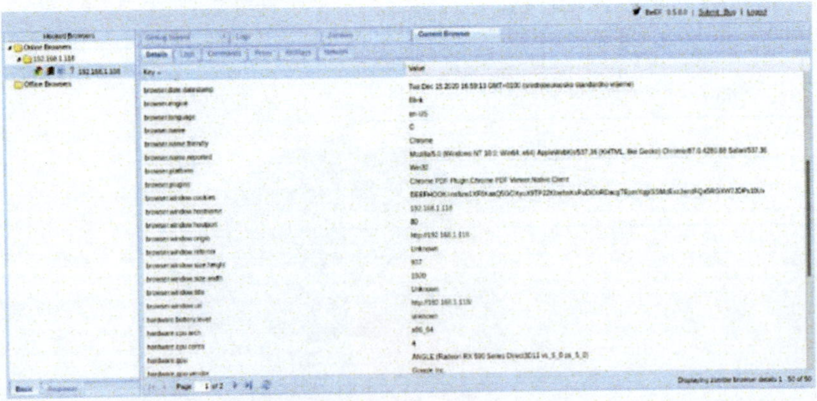

Fig. 5 Details of connected browser

simulation. Apache server was started in order for the website to be accessible on the local network.

4.1 Examples of Attacks Using a BeEF Framework

Immediately after BeEF is started, the Linux terminal opens with the message that the BeEF server is started and with an example of the Hook.js script must be implemented and that is *<script src=http://<IP>:3000/hook.js></script>*, where "IP" needs to be replaced with the IP address of an attacker's computer. After that, that HTML tag with path is implemented into the HTML code of the website.

After opening the malicious website in the attacked browser, the response is immediately seen in the user interface of BeEF. In other words, the attacked browser comes up under "Online browsers" folder. Figure 5 shows some of the information that can be seen under the Details tab of a connected browser and as it can be seen, some valuable information can be gathered just by analyzing that tab.

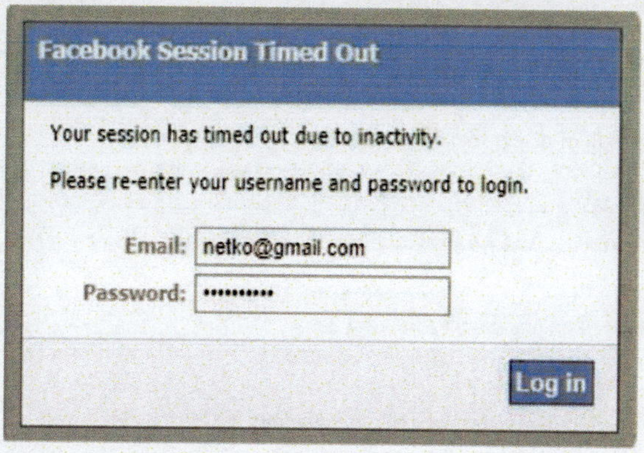

Fig. 6 A fake Facebook pop-up window

BeEF has many different attacks that can be executed to gather more sensitive information from the victim. In this research, five examples of the attacks were tested. Those are:

- Pretty theft
- Google phishing
- Detect social networks
- Get page HTML
- Remote access using BeEF and Metasploit

Pretty theft is an attack that can be executed in various forms and its primary purpose is to gather data like username and password for a social network. For this research, the attack is executed in the form of a Facebook log-in pop-up window. The result of the attacked browser is shown in Fig. 6.

After clicking the log-in button, the entered email and password can be seen by the attacker in the BeEF user interface. A Google phishing attack has a similar purpose to pretty theft, but as its name suggests, it is intended for stealing Google log-in credentials. The important thing about this attack is that it sends the email and password to the attacker and redirects the victim's browser to the real Gmail interface, making Google phishing harder to notice as an attack.

The third tested attack is detect social networks and as a result it gives information about social networks the attacked user is logged in. This attack can be used in combination of other attacks to make them more effective, for example, the previous pretty theft or Google phishing. The result of an attack is shown in Fig. 7.

Get page HTML is the fourth tested attack and the purpose of this attack is to gather the HTML code of the currently opened website. After executing the attack, the BeEF user interface is shown the HTML code of the malicious website created for the simulation because the website is currently opened in the attacked browser.

Fig. 7 Result of detect social networks

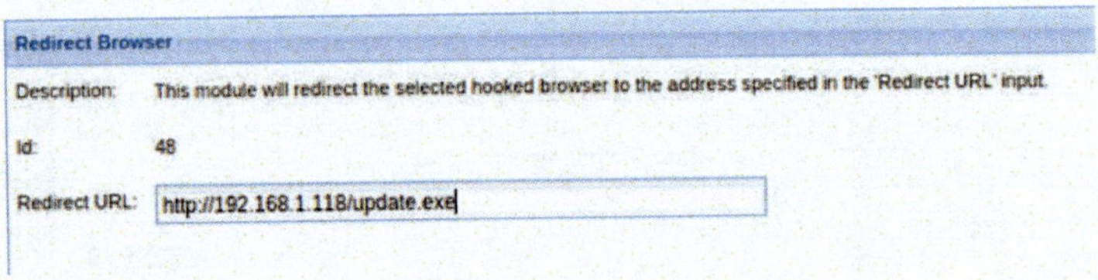

Fig. 8 Redirecting the browser to the malicious payload

In the real-life scenario, this attack could be used for analyzing the code for further implementation of malicious code or scripts.

4.2 *Remote Access Using BeEF and Metasploit*

For more significant results of the attack, sometimes it is necessary to combine different tools. In this example, Metasploit was used in combination with BeEF to gain remote access to the attacked computer. Metasploit is a framework used for penetration testing and ethical hacking. It includes interfaces and modules that can be used for hacking and creating malicious payloads [20].

The malicious payload needs to be created and sent to the attacked computer to gain remote access to the computer. For creating a payload for this simulation, MSFvenom was used. Many different payloads can be generated with this tool, and in this case, *reverse_tcp* was specified. On attacking the computer, the listener needs to be set up. The listener is active after the start and notifies the started session the moment the malicious file is started. Therefore, it is crucial to align data like LPORT, LHOST, and specified payload when creating the payload and setting up the listener to know "what to listen for." For more straightforward navigation through directories, once the access is obtained, Meterpreter is used.

When the Metasploit part is ready, BeEF is used to deliver the malicious payload to the attacked computer. It can be done in various ways, but in this research, Redirect Browser attack was used. To redirect the browser to the malicious payload, the attacker needs to specify the path to the file in the BeEF user interface. In this case, it was the IP address of an attacking computer following the "/Update.exe." Update.exe is the name given to the malicious file to be more convincing. Figure 8 shows the redirecting of the browser from BeEF.

When redirecting of the browser is executed, the download request appears on the attacked browser. If the attacked user is not well informed about the dangers on the Internet, he/she could download and start the payload without knowing the

```
Terminate channel 1? [y/N]  y
meterpreter > ls
Listing: C:\Users\DomVirtual\Documents
======================================

Mode              Size  Type  Last modified              Name
----              ----  ----  -------------              ----
40777/rwxrwxrwx   0     dir   2021-01-23 14:45:09 +0000  My Music
40777/rwxrwxrwx   0     dir   2021-01-23 14:45:09 +0000  My Pictures
40777/rwxrwxrwx   0     dir   2021-01-23 14:45:09 +0000  My Videos
40777/rwxrwxrwx   0     dir   2021-01-24 12:49:08 +0000  datoteke
100666/rw-rw-rw-  402   fil   2021-01-23 14:46:18 +0000  desktop.ini
40777/rwxrwxrwx   0     dir   2021-01-24 10:45:40 +0000  exclusion folder
40777/rwxrwxrwx   0     dir   2021-01-24 10:45:47 +0000  fotografije

meterpreter >
```

Fig. 9 Files and folders on the attacked computer

consequences. After the payload is executed, there is a notification on the attacking computer that the session is started. From this point, the attacker has access and can navigate through all directories on the victim's computer. Figure 9 shows the list of files and folders in the current directory.

Various commands can be executed to gather more information from the attacked side. For example, the sample image was successfully fetched from attacked computer to the attacking computer with the download command. Furthermore, with the edit command, the attacker can edit documents, and in this editing, a text document was tested and resulted in modification of the text file that was located on attacked computer.

5 Synthesis of Results and Proposal of Protection Guidelines

Simulation conducted for this research was described through the previous chapter. Five attacks were simulated and each of them can be conducted as a separate attack or in combination with others, depending on the wanted goal. Each of the conducted attacks is stated in Table 1. Also, for each attack in Table 1 there are results of the attack, and it is a potential consequence for the attacked user.

As it can be seen from the results of the simulation pointed out in the table, the simulated attacks succeeded. Based on this, it can be concluded that additional commitment of an individual user is required to protect against XSS attacks adequately. Therefore, protection proposals were tested and analyzed in this research.

The first method is to block JavaScript code in the web browser itself. This method does not require any additional tools because most web browsers available today have this option implemented in their settings. The browser used in this example is Google Chrome, and in the browser's settings execution of JavaScript was turned off. With that setting, the browser was not connected to BeEF when opening the malicious web page, and the message "JavaScript was blocked on this page" showed in the browser. That means the method worked, but it has a downside.

Table 1 XSS attack simulation results

Attack	Result	Consequences for the user
Pretty theft	• Facebook log-in pop-up window shows • User's input visible to the attacker	• The attacker has access to log-in information • Possibility to log in to one of the social networks • Possible compromise of log-in to other services if the user uses the same log-in information
Google phishing	• Fake Gmail log-in interface opens • Browser redirected to the real Gmail interface • User's input visible to the attacker	• An attacker has access to Gmail log-in credentials • Ability to sign in and read the complete content of Gmail • Possible compromise of log-in to other services if the user uses the same log-in information
Detect social networks	• Information is visible in the BeEF interface if the user is logged in to one of the offered social networks	• An attacker can use information about what social networks the attacked user is using • Based on the results of this command, the attacked user is more vulnerable to further attacks
Get page HTML	• The HTML code of the currently open page is displayed in the BeEF interface	• An attacker can analyze more vulnerabilities from the website code • Ability to implement malicious scripts based on analyzed code
Redirecting the browser to the malicious payload	• Browser redirected to download a malicious file • Gains remote access to the attacked computer	• User's computer is compromised by downloading the file • Information theft • Execution of commands

This will block all the JavaScript on all the pages, which means good ones also. So most of the pages will lose its dynamic elements, and some will even be unusable. However, the user can define which pages the blockade refers to, so it is possible to determine a page that the user suspects could contain dangerous scripts and thus reduce the negative aspects of this method.

The second method tested requires the installation of the uMatrix tool. It is a simple tool available on GitHub, but it also comes as an extension for Google Chrome and is available in the Chrome web Store. It is quite easy to install and use, so it is suitable for most users. After activating uMatrix, the browser did not appear as connected in BeEF interface. Therefore it is concluded that uMatrix secured the attacked browser. Also, in the software's log, BeEF cookies were listed.

The third tested tool differs from the ones shown before, and it is called XSpear. It is an open-source tool for scanning for XSS vulnerabilities. It is more suitable for detecting reflected XSS vulnerabilities, so it was vulnerable and publicly available for such purposes; Acunetix website shows an example of how the tool works. The tool listed found vulnerabilities in the output of the command and marked the severity of each vulnerability. XSpear proved itself to be great, i.e., for web developers to test their website for XSS vulnerabilities.

The last tested method is using Burp Suite. It is a software designed for network security, and it is available on Windows, Linux, and macOS. Before using it to scan for malicious payload web browser was set up to use Burp Suite as a proxy, so all data passed through Burp while browsing. After Burp Suite was started, the malicious website was opened, but the browser did not appear in the BeEF interface as connected, which means that the defense method worked. Also, in the log of the Burp Suite there was a message that the "BEEFHOOK" cookie was found.

Each defense differs in the way of implementation and functionality and each has its own advantages and disadvantages. During the testing, the listed characteristics were monitored and are listed in Table 2.

Each method fulfilled its purpose during testing, but the characteristics are not the same. Blocking JavaScript is the simplest method, but it has a big drawback, and that is with its implementation, websites and applications lose their dynamic elements. The implementation of the uMatrix tool has proven to be the most suitable method for the end user. It is installed as an extension in a web browser, so it can be said that the implementation is also simple. The tool successfully blocked a malicious script, thus preventing browser compromise, while other, non-harmful JavaScript elements on websites are not blocked, thus enabling unhindered use and visiting of web pages and applications.

6 Conclusion

With the increase in the use of the Internet for everyday activities comes an increase in the risk of cyberattacks. Any user of the Internet and services provided via the Internet can be the target of an attacker. An analysis of previous research shows that the prevalence of XSS attacks is high, and the very fact that OWASP included them in the top ten list speaks of the seriousness of these attacks. Many research has been conducted about XSS attacks and proposed protection guidelines, which help to raise awareness about these attacks.

The research shows a simulation of XSS attacks using the BeEF framework. This framework was chosen because of its availability and can therefore be used by anyone. The fourth chapter shows examples of the tool's possibilities by which the attacker can get certain information that he/she can use for further malicious actions. XSS attacks often use multiple tools to give the attacker more opportunities to exploit browser vulnerabilities, so this research also shows an example of using BeEF in combination with the Metasploit framework. This example shows how

Table 2 Protection proposal analysis

Defense method	Implementation	Functionality	Advantages	Disadvantages
JavaScript blocking	• Activation in browser settings	• Blocks execution of JavaScript when opening websites	• Simple to implement • No additional installation is required • Available on most browsers	• Blocks all JavaScript, including the non-malicious ones
uMatrix	• Installation as browser plug-in	• Blocks execution of malicious JavaScript • Saves scanning results in the log	• Simple to implement • Blocks only JavaScript elements that are recognized as risky • Open-source code	• Possible incorrect risk assessment • Modification could be complicated to some users
XSpear	• Installation on Linux OS	• Detects XSS vulnerabilities in a given website	• Good at detecting reflected XSS vulnerabilities • Good tool for web developers	• Did not detect stored XSS in tested malicious website • Available only on Linux
Burp Suite	• Installation on OS • Configuration in web browser	• Scans all data passing through the browser • Blocks malicious scripts and stores in log	• Software has a lot of functionalities • Available on multiple OS • Reliable	• Free version does not include all of the functionalities • Most complicated to use

an XSS attack can be used to access a computer and all the data on it remotely. By analyzing the simulation results and the fact that the simulated attacks were successful, it was concluded that using the available tools to perform attacks, an attacker can cause severe damage to the attacked user. This once again confirms the necessity of every user to take protective measures.

In the last section of the chapter, the research results are synthesized and presented through a table. For each simulated attack, the potential consequences it may have for the attacked user are listed. Based on the analysis of the results, protection guidelines have also been proposed. Of the proposed methods, by the best ratio of functionality and simplicity uMatrix stands out the most. It was tested by opening a website made for this simulation and it immediately blocked the malicious script. As a result, BeEF could not connect to the browser and it was not possible to execute commands and collect data.

This research gives solid guidelines to improve the resilience of end users on XSS attacks, which can prevent the further negative impact of such attacks on Industry

4.0 systems. In future research, we plan to implement those guidelines in real Industry 4.0 environment and measure its effectiveness and resilience improvement of such environment.

References

1. I. Cvitic, D. Perakovic, M. Perisa, M. Botica, Definition of the IoT Device Classes Based on Network Traffic Flow Features, in *EAI/Springer Innovations in Communication and Computing*, ed. by L. Knapcikova, M. Balog, D. Perakovic, M. Perisa, (Springer, Cham, 2020), pp. 1–17. https://doi.org/10.1007/978-3-030-34272-2_1
2. D. Peraković, M. Periša, I. Cvitić, P. Zorić, *Information and Communication Technologies for the Society 5.0 Environment*, in XXXVIII Simp o novim Tehnol u poštanskom i Telekomun saobraćaju – POSTEL 2020 [Internet], University of Belgrade, Faculty of Transport and Traffic Engineering, Belgrade, Serbia (2020), pp. 203–212. https://ebooks.sf.bg.ac.rs/index.php/FTTE/catalog/view/15/45/139-1
3. A. Sołtysik-Piorunkiewicz, M. Krysiak, The cyber threats analysis for web applications security in industry 4.0. Stud Comput Intell, 127–141 (2020). https://doi.org/10.1007/978-3-030-40417-8_8
4. B.B. Gupta, A. Tewari, I. Cvitić, D. Peraković, X. Chang, Artificial intelligence empowered emails classifier for Internet of Things based systems in industry 4.0. Wirel. Netw. **28**, 0123456789 (2021)
5. I. Cvitić, D. Peraković, M. Periša, S. Husnjak, An overview of distributed denial of service traffic detection approaches. PROMET Traffic Trans. **31**, 453–464 (2019). https://traffic.fpz.hr/index.php/PROMTT/article/view/3082
6. B.B. Gupta, P. Chaudhary, S. Gupta, Designing a XSS defensive framework for web servers deployed in the existing smart city infrastructure. J. Organ. End User Comput. **32**, 85–111 (2020)
7. I. Yusof, PKA. Preventing Persistent Cross-Site Scripting (XSS) Attack By Applying Pattern Filtering Approach (2015)
8. M. Johns, B. Engelmann JP. XSSDS: Server-side Detection of Cross-site Scripting Attacks (2008)
9. P. Wurzinger, C. Platzer, C. Ludl, E. Kirda, KC. SWAP: Mitigating XSS Attacks using a Reverse Proxy. Tehničko sveučilište Beč (2009)
10. Acunetix (2021). Cross-site Scripting (XSS). Retrieved April 16, 2021, from https://www.acunetix.com/websitesecurity/cross-site-scripting/.
11. S. Forge, J. Grossman, R. Hansen AR., *Cross Site Scripting Attacks: XSS Exploits and Defense* (Syngress Publishing, Burlington, MA, 2007)
12. B.B. Gupta, P. Chaudhary, *Cross-Site Scripting Attacks Classification, Attack, and Countermeasures* (CRC Press, 2020)
13. I. Yusof, A.K. Pathan, *Preventing Persistent Cross-Site Scripting (XSS) Attack by Applying Pattern Filtering Approach*, in The 5th International Conference on Information and Communication Technology for The Muslim World (ICT4M) (2014), pp. 1–6. https://doi.org/10.1109/ICT4M.2014.7020628
14. M. Johns, B. Engelmann, J. Posegga, *XSSDS: Server-Side Detection of Cross-Site Scripting Attacks*, in 2008 Annual Computer Security Applications Conference (ACSAC) (2008), pp. 335–344. https://doi.org/10.1109/ACSAC.2008.36
15. P. Wurzinger, C. Platzer, C. Ludl, K. Engin, C. Kruegel, SWAP: Mitigating XSS attacks using a reverse proxy (2009), pp. 33–39. https://doi.org/10.1109/IWSESS.2009.5068456
16. Acunetix, Cross-site Scripting (XSS) (2021), https://www.acunetix.com/websitesecurity/cross-site-scripting/. Accessed 16 Apr 2021

17. J. Grossman, R. Hansen, P.D. Petkov, A. Ragger, S. Fogie, *XSS Attacks Cross Site Scripting Exploits and Defense*, 1st edn. (Amorette Pedersen, Burlington, 2007)
18. B. Gupta, B., Chaudhary, P., *Cross-Site Scripting Attacks Classification, Attack and Countermeasures*, 1st edn. (CRC Press, Boca Raton, 2020)
19. Acunetix, What Is Persistent XSS (2021), https://www.acunetix.com/blog/articles/persistent-xss/. Accessed 18 Apr 2021
20. OWASP, DOM Based XSS (2021), https://owasp.org/www-community/attacks/DOM_Based_XSS. Accessed 18 Apr 2021
21. WhiteHat Security, 2017 Application Security Statistics Report (2021), https://info.whitehatsec.com/Content-2017-Stats-Report-LP.html?utm_source=website&utm_medium=Website-Content-2017-StatsReport. Accessed 18 Apr 2021

Education and Its Role in Risk Management in the Context of Industry 4.0

Juraj Sinay, Aleš Bernatík, and Vendula Laciok

1 Introduction

The term Industry 4.0 means a method of technology management that already exists in some areas of industrial production, where machines and products communicate with each other and organize the individual steps in the production process autonomously, i.e., themselves.

Customers increasingly inform themselves of the products or services they want to procure/use and expect maximum flexibility from manufacturers. They do not want to buy products with the features offered by production organizations. They themselves want to define the characteristics of the product they will pay for. This method of strong individualization requires a high degree of flexibility in series production by involving both customers and business partners in the chain of creating the final product characteristics based on partnerships that come from different areas of industry.

The Industry 4.0 strategy includes in particular:

- Linking of production with information and communication technologies
- Possibilities to implement customer requirements directly through data mobility (communication) from machines and equipment

J. Sinay (✉)
Faculty of Mechanical Engineering, Technical University of Košice, Košice, Slovakia

Faculty of Safety Engineering, VSB Technical University of Ostrava, Ostrava, Czech Republic
e-mail: juraj.sinay@tuke.sk; ales.bernatik@vsb.cz; vendula.laciok@vsb.cz

A. Bernatík · V. Laciok
Faculty of Safety Engineering, VSB Technical University of Ostrava, Ostrava, Czech Republic
e-mail: juraj.sinay@tuke.sk; ales.bernatik@vsb.cz; vendula.laciok@vsb.cz

L. Knapčíková, D. Peraković (eds.), *6th EAI International Conference on Management of Manufacturing Systems*, EAI/Springer Innovations in Communication and Computing, https://doi.org/10.1007/978-3-030-96314-9_4

- Communication between the components of production processes with each other
- Autonomous data acquisition and processing on a vertical and horizontal level
- Decentralized management
- Realization of production through communication between semifinished products and machinery and handling equipment

The fourth industrial revolution will change the nature of the labor market. Some jobs will be replaced by robots as part of automated production. It is possible to assume that the number of jobs will be reduced, where the requirements for the qualification of employees will be reduced. Modern technologies and their deployment in production processes will require the qualification of employees for new job positions.

Manufacturing technologies as part of Industry 4.0 will place new demands on risk reduction both in the work environment and in the production process. Machines and machine systems will communicate and interact with man in the man-machine-environment system using sensors and artificial intelligence. The machines will be equipped with control systems that will be able to predict the occurrence of a failure and then control their maintenance systems. With the help of sensors, it will be possible to minimize the dangers due to collisions between machines and between man and machine. In this context, new risks are emerging associated with the disruption of IT management systems from the external environment.

For the worker who will work with intelligent machines, the term Operator 4.0 is often used, which can be characterized as an intelligent (smart) and capable expert who not only cooperates with robots but also performs work using cyber-physical systems, adaptive automation for the purpose of cooperation within corporate governance structures. Operator 4.0 will thus have control over the work process and technologies that support the use of human capabilities. It becomes part of the man-machine-environment system and thus a source of risk.

The activity of Operator 4.0 can be oriented both monothematically, i.e., on one type of activity, and on several activities, if required by the nature of the automated production technology.

Operator 4.0's work operations must be part of risk analyses, while the failure of the human factor can only be described with exact approximation by exact algorithms [1, 2]. Therefore, it is important to look for ways to increase the reliability of its conduct. Acquiring knowledge as a prerequisite for acquiring a skill is a strategic prerequisite for the competence of Operator 4.0. Applications of Industry 4.0 systems in industry are currently at the stage of their deployment in some technologies, which are characterized by a high degree of automation. The training of experts in university study programs does not have defined curricula/study programs. In cooperation with industry, the contents of programs and professional courses will be modified.

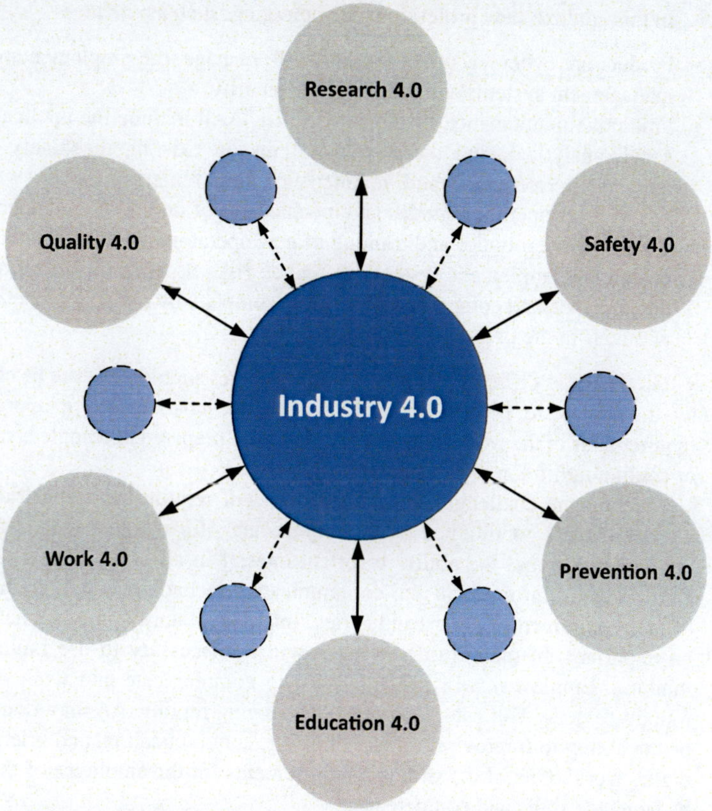

Fig. 1 New challenges within Industry 4.0

1.1 *Impact on Education*

Industry 4.0 brings technical and social development to the current world of work. One of the strategic tasks in the near future is to link adequate education, currently referred to as Education 4.0, based on the results of applied research and innovation, to Safety 4.0, Prevention 4.0, Quality 4.0, and the like (Fig. 1) [1].

The requirements for specialists at different levels of education for work performed within Industry 4.0 will be based on the nature of their work. New areas of knowledge and competence can include:

- Principles of operation of the Internet
- Its use in selected professions
- Applications of new technologies in various industries
- Knowledge of the location and tasks of all components in the Industry 4.0 chain (e.g., sensors, IoT networks, clouds, ICT control systems)

In this context, new professions are emerging, such as:

- Production cyber security expert—will manage the implementation of risk management systems and ensure cyber security.
- Predictive maintenance systems specialist—will include the application of seasonal, analytical, and diagnostic systems on existing machinery to monitor machine performance and maintenance requirements. He/she will have an overview of operating modes and mechanisms of failure of machinery.
- Collaborative robotics and training of new operators in their use specialist—will design and implement systems to reduce risks at work, in order to ensure the performance and competitiveness of the company by replacing repetitive manual activities in the production process.

Graduates of different types of training courses should in future be characterized by the ability to be flexible and creative, to have an active approach to the requirements of the work environment, and to be prepared to complete various forms of continuing/lifelong learning [2].

Meeting the challenges of Industry 4.0 will require the acquisition of digital literacy. This is conditioned by the requirements for working with a large amount of information and the ability to orient oneself in all areas of new technologies. Knowledge of information and communication technologies will gradually be part of the requirements of most industries. Interdisciplinarity and technical skills will be important. In many professions, it will be necessary to use multidisciplinary thinking. Employees will need to learn to communicate and work directly with robotic devices. The nature of work is changing rapidly. An important aspect will be the ability to improvise, logical thinking, communication, and orientation in the digital world. One of the strategic requirements for the employee of the future will be language skills and competence.

1.2 Consequences of New Working Conditions

It is not possible to generalize the psychological consequences of new working conditions, but also of new systems, which will also include robots. They will have an individual character and will depend on the characteristics of the person.

Technological progress will lead to a further reduction of physical strain and the risks due to mechanical hazards will be reduced. This will result in the improvement of the hygienic parameters of the working environment as well as the improvement of the quality of personal protective equipment. Health and safety legislation will reflect changing working conditions. Mental and physical threats to a person will depend on the changing characteristics of the effects of stress due to the performance of work, but also on the speed of technological change and thus on the ability of employees to adapt to them.

Emerging risks arise as a result of work activities that are the result of the application of new technologies in industrial enterprises. Their minimization leads to ensuring the physical and mental health of employees.

Technological development within Industry 4.0 will lead to a reduced risk of working conditions, which by their nature of work are burdened by high risk. Effective measures for their reduction can include, in particular, effective preventive measures in the stage of product design as well as in workplace designing.

The most common cause of accidents at work is an incorrectly estimated risk. This is also associated with insufficient training of experts in risk management, which in the future will require paying increased attention to their education and effective acquisition of competencies in the field of safety and health at work as well as the safety of machines and machine systems.

To ensure the conditions for the training of experts in the field of risk management, it is important to know the requirements of partners from practice in various areas of industrial activities as well as public and social life. One of the most effective methods for obtaining information is experiments carried out in the form of questionnaire surveys.

2 Results of Individual Surveys

In connection with obtaining information on the requirements of employers, there is a need to identify their views, experiences, and needs of the target group, i.e., employers, occupational safety inspectors, employee/trade union representatives, as well as students of relevant study programs. Attention was paid mainly to the characteristic areas of Industry 4.0, which can include digitization, automation, robotics, and their consequences for changing the nature of work professions. The method of quantitative field research was chosen for the implementation of the experiment, while the principles of compliance with the conditions arising from the GDPR rules were applied [3].

The questionnaire was devoted to the characterization of business entities in the context of Industry 4.0. Its structure is shown in [4]. The aim of the experiment was to analyze the current state of companies with a focus on cooperation with educational institutions, including the formulation of requirements for graduates. The research included companies that have Industry 4.0 technologies in place. In this case, it was a matter of obtaining feedback on the possible transformation of study fields for the requirements arising from the requirements of Industry 4.0. The questionnaire contained 11 questions.

2.1 Evaluation of Hypotheses

The data obtained by questionnaire research were tested using four hypotheses at the 5% level of significance ($\alpha = 0.05$). The Good Match Test (Pearson's chi-square test) was used for testing and is used for frequency comparison data. The general formulation of the null and alternative hypotheses is:

H0: frequencies are the same.
HA: frequencies differ significantly.

The test statistics is evaluated using a formula:

$$\chi^2 = \sum_{i=1}^{n} \frac{(O_\mathrm{i} - E_\mathrm{i})^2}{E_\mathrm{i}} \quad (1)$$

Where:

O_i = measured (observed) frequency of the i-th group
E_i = estimated (expected) frequency of the i-th group
n = number of groups
χ^2 = value of the tested chi-square distribution statistic

The test was performed using the CHIDIST function when the level of significance (p-value) was reached. This value was then compared with the selected level of significance (α). The null hypothesis was rejected in favor of the alternative hypothesis if the achieved level of significance was lower than the chosen level of significance. The calculations were performed using Microsoft Excel 2007 software.

Hypothesis No. 1: Employers prefer graduates of the Czech Technical University in Prague, in contrast to VŠB-TU Ostrava.

The achieved significance level (0.4142) was higher than the chosen significance level (0.05); therefore the null hypothesis will not be rejected in favor of the alternative hypothesis. The hypothesis that "Employers prefer CTU graduates, in contrast to VŠB-TUO" was not confirmed at the 5% level of significance.

Hypothesis No. 2: Employers do not care at which school their future employees studied.

The achieved level of significance ($1.49 * 10^{-7}$) is lower than the selected level of significance (0.05). The test showed that the frequency varies significantly. It has been shown that the abundance of n_N (68) is significantly different; it is higher compared to the abundance of n_Z (19). This proves that the hypothesis that "Employers do not care at which school their future employees studied" is confirmed at the 5% level of significance.

Hypothesis No. 3: After graduating from university, students' knowledge is insufficient for the needs of employers.

The achieved level of significance (0.0243) was lower than the selected level of significance (0.05). The test showed that the abundance of n_N (33) differs significantly compared to the abundance of n_D (54). This confirmed that the correctness of the hypothesis was not confirmed; that is, it was rejected at the 5% level of significance.

Hypothesis No. 4: More than half of the employers contacted, who did not cooperate with universities in organizing student internships, would welcome this opportunity in the future.

The significance level achieved (0.1444) was higher than the chosen significance level (0.05); therefore the null hypothesis will not be rejected in favor of the alternative option. The hypothesis that "More than half of the addressed employers who did not cooperate with universities in organizing student internships would welcome this opportunity in the future." could not be confirmed at the 5% level of significance.

The results of the questionnaire experiment made it possible to define the following conclusions:

- Recommendations for the cooperation of universities with companies of industrial practice already during their studies and not to wait for the completion of studies.
- Graduates' knowledge of new technologies needs to be increased once they have entered practice. Study programs must be aimed at acquiring skills for information processing so that graduates are able to design security structures for projects.

The following recommendations have emerged for risk management systems:

- Promote the introduction of integrated safety principles in industrial companies as part of complex management systems.
- Support the introduction of digitization in companies with the support of the company's top management.
- It was confirmed that the employees are adequately prepared for the implementation of Industry 4.0 strategy, despite the fact that in real industrial conditions, the Industry 4.0 strategy is only partially implemented.

3 Impact of the Covid-19 Pandemic on Education in the Context of Industry 4.0

Considering the original strategy of Industry 4.0, which was based on the philosophy "Supply products according to the wishes of the market/customer," the application of ICT in all stages of logistics processes focused on the influence of the human factor in production parts of the production stage is minimized. On the other hand, the share of human labor is increasing in design activities, with the application of software products in the design of complex automated systems. The application of the principles of Industry 4.0 is currently implemented mainly in production/distribution technologies characterized by a high degree of automation using robotic and handling complexes. Within the interconnection of individual components of the system, efforts are made to exclude the human factor not only in the primary stages but also at their interfaces [5].

These are mainly:

- Sensors on machinery—machine-sensor interface
- Data mobility from the machine-sensor primary location to the operational center processing of the obtained information—onboard computer (board computer)
- Data mobility from the board computer to the cloud
- Cloud data processing (application of Big Data methodology)
- Reverse mobility of data from the cloud to the place of the actuator as an active member on the machine, a device that performs the necessary operation on the technological device

The place of the human factor in Industry 4.0 is above all:

- In the field of ICT design for system management in the context of the interconnection of individual parts
- When verifying the information obtained
- When performing repairs and maintenance activities

After considering the above factors, it can be assumed that risk management systems in the context of Industry 4.0 will be modified primarily to minimize them in areas where there is a significant place for man.

During the design activity stage and verification of the information used, the characteristics of the threats and thus of the risks are changing. New types of risks are emerging based on a person's mental activity and his/her mental/psychological load (e.g., stress, ability to concentrate—mental health).

This finding also determines the choice of means to minimize these types of risks. It is possible to assume that in the future, in the framework of risk prevention as part of education and acquisition of competencies, methods will be used, e.g., mental training, mental regeneration, physical activities, and application of communication techniques.

It can be stated that the impact of a pandemic, e.g., Covid-19, will not require changes in the implementation of risk management systems as part of integrated management systems in companies. The reason is the reduction of the share of the human factor and thus the consequences of the impact of the pandemic on the operation of various types of industrial technologies.

A strategic change in the incidence of threats in the context of Industry 4.0 will place significant emphasis on changing people's education and training. The requirements for ways and methods for acquiring knowledge, skills, and competence will change. Emphasis will not be placed on the knowledge acquired in the form of full-time forms of education (although it is not possible to omit this form), but mainly on the combination of the connection between education and gaining practical experience [6].

As a result of these changes, it can be expected that part of a person's preparation for a permit will not be possible only in the form of online education. In these scenarios, their changes will need to be made through training programs using specific areas of industrial activity, which will require intensive cooperation between educational institutions and industry, with the active participation of supervisory authorities from the structures of national authorities for occupational safety and health management.

The Covid-19 pandemic accelerated the development and application of the implementation of the Industry 4.0 strategy into industrial practice. In professional circles, it was assumed that Industry 4.0 will be the cause of the increase in unemployment. All the implications of the Covid-19 pandemic have been shown to have a positive effect on changing people's thinking.

4 Recommendations for the Implementation of Industry 4.0

Industry, academic community, and regulatory authorities should understand and accept the perception of risk minimization and thus the importance of prevention within society, even if these concerns are based on a lack of knowledge and competence. Public awareness and involvement in risk management are important, and the aim must be to involve the public in the role of an active participant. This fact will require the preparation of educational programs at all levels of education, including lifelong learning.

This will highlight the share of acquired knowledge as a strategic condition for the acquisition of competencies. In these activities, it is important to use the experience of partners from industrial practice at all stages of the training of qualified experts in the field of risk analysis in the technologies of Industry 4.0. Their feedback can be one of the decisive criteria for the quality of education in the framework of study programs focused on the safety of new techniques and technologies.

4.1 University Education

The academic community is an important center for the acquisition of knowledge, skills, and partly also competence in order to contribute to the creation of a safe life and thus to the reduction of risks in society. The method of systematic vocational training at universities must be based on three pillars of education, research, and cooperation with practice, while contacts with employers must create, in feedback, conditions for preparation and implementation of a comprehensive approach to training professionals for the introduction of Industry 4.0 into integrated company management systems.

Effective implementation of risk management processes for product safety and technological processes is conditioned by gaining practical experience. Students must gain the belief that safety as a property of products, production, and logistics technologies in the context of process safety must be part of all educational processes. Conditions in laboratories at universities should create conditions for simulating processes in real conditions of practice.

As part of the study programs, universities should not only provide students with technical skills but also enable them to acquire the principles of safety culture throughout their studies in all fields of study, especially in the field of technology [7]. In educational process, measures are proposed, which will make it possible to contribute to increasing the employability of graduates in the conditions of Industry 4.0.

5 Conclusion

Industry 4.0 strategy is changing the working environment and conditions today. Therefore, new content and methods must be integrated into educational processes in all their forms when preparing human potential. Topics such as cyber-physical systems, cyber security, robotics, and social media play a crucial role in networked production processes [8].

The principles arising from the application of the Industry 4.0 strategy must be integrated into the study programs of secondary vocational schools and universities, with the principles of this strategy having to be emphasized in basic schools on simple examples.

A fundamental change in the value chain of a product and its life cycle due to the interconnection of the real and virtual world requires learning based on the need to acquire appropriate competencies. One of the important attributes of "education of the future" must be the requirement of the existence of minimal risks in all areas of social and public life as part of preventive measures.

The qualification of employees is the most important challenge today for companies in various industries. In the following years, several new work areas will emerge in companies, e.g., in the field of manual operation of machines as well as work on the belt, and other classic ones will gradually disappear. Simple work manipulations will be performed by robots.

However, new occupations will emerge that will require greater competence in the use of IT techniques as well as readiness for lifelong learning. New risks will arise in connection with this. The result will be a change or adjustment of study programs and educational courses. They will include procedures and methodologies for acquiring skills and competences for all areas of risk management systems in cooperation with industry. Attention will be paid to Prevention 4.0 as part of integrated business management systems.

The Association of the German Engineering Industry (VDMA) states in "Industry 4.0-Qualifizierung 2025" [9] study that the biggest challenge for ensuring the competitiveness of the German economy is in the skilled workforce. Without a highly skilled workforce, it will not be possible to gain a competitive advantage in a globalized economy.

At the same time, the focus of Strategy 4.0 is and will always be human. Therefore, it is important to create conditions for its safety both in the workplace and in everyday life.

Acknowledgements This contribution is the result of the project's implementation: APVV No. 19-0367 "Framework of the Integrated Process Safety Management Approach for the Intelligent Enterprise." The authors also gratefully acknowledge the financial support of the project "Potential Impacts of Industry 4.0 on Operators in 3.0 Jobs and Tertiary Education in Accordance with Safety Engineering," identification code TL01000470.

References

1. SWOT analýza Národní iniciativy Průmysl 4.0 - Znalostní systém prevence rizik v BOZP. Úvod - Znalostní systém prevence rizik v BOZP [online] (2016), https://zsbozp.vubp.cz/pracovni-prostredi/odvetvi/prumysl-4-0/468-swot-analyza-narodni-iniciativy-prumysl-4-0. Accessed Jan 15 2021
2. Průmysl 4.0 - úvod do problematiky - Znalostní systém prevence rizik v BOZP. Úvod - Znalostní systém prevence rizik v BOZP (2016), Dostupné z, https://zsbozp.vubp.cz/pracovni-prostredi/odvetvi/prumysl-4-0/464-prumysl-4-0-uvod-do-problematiky. Accessed Jan 15 2021
3. J. Sinay, Z. Kotianová, *Automotive Industry in the context of Industry 4.0 Strategy/Automobilový priemysel v kontexte Stratégie Priemysel 4.0 In: XIII. konferencia Bezpečnost a ochrana zdraví při práci*, BPBI, Ostrava, ČR, str. 73 až 76; ISBN 978-80-7385-202-3
4. Idnes.cz, Evropská komise: Česko – země robotům zaslíbená (2020), https://sdeleni.idnes.cz/zpravy/evropska-komise-cesko-zeme-robotum-zaslibena.A201118_131157_zpr_sdeleni_hradr/Českárepublika
5. IBA Global Employment Institute Artificial Intelligence and Robotics and Their Impact on the Workplace, Prepared by: Gerlind Wisskirchen and team (2017)
6. Dopady Průmyslu 4.0 na trh práce v ČR, Národní observatoř zaměstnanosti a vzdělávání Národní vzdělávací fond, o.p.s., Česká republika (2017)
7. Potenciální dopady Průmyslu 4.0 na pracovní pozice Operátorů 3.0 a terciální vzdělávání s ohledem na bezpečnostní inženýrství; Technologicka agentura České republiky, projekt: TL01000470; Vysoká škola bánska –Technická univerzita Ostrava, Fakulta bezpečnostního inženírství/Česká republika (2021)
8. Tendy v robotizaci a automatizaco. Znalostní systém prevence rizik v BOZP, https://zsbozp.vubp.cz/kalendar-akci/150-trendy-v-robotizaci-a-automatizaci?date=2018-01-31-00-00. Accessed 15 Jan 2021
9. Industrie 4.0-Qualifizierung 2025 (2016), https://www.ihre-industrie.de/fileadmin_upload/2016-05-18_VDMA_Bildung; BRD

Using Quality Management Tools to Analyze the Error Rate in the Production Process

Miriam Andrejiova, Miriama Pinosova, and Ervin Lumnitzer

1 Introduction

Quality has become one of the most important factors in consumer choice between competing products and services [1]. This ensures that it is a controlled process that includes people, a system of production, or service provision, and supports tools and methods, so it can be measured using quality management standards from the EN ISO 9000:2000 series Quality Management Systems-Fundamentals and Vocabulary (definitions) (EN ISO 9000, EN ISO 9001, and EN ISO 9004). Under ISO 9000, quality is actually the sum of all product's characteristics and features related to the ability to meet specific requirements [2] that are usually determined by customers and their demand for quality in a product or service is constantly increasing.

According to Kumar Sharma and Gopal Sharma [3], global competition forces small and medium-sized enterprises to increase their competitiveness by increasing the performance of their production. They must pay attention to the reliability of their production processes and also their commitment to quality management procedures. Effective improvement of quality leads to increased productivity and reduction of production costs and thereby strengthens the position of the organization in the market. Quality is one of the tools of competitiveness and it is an integral part of the overall business strategy of organizations.

Quality management is a targeted activity of the organization through quality management systems (QMS), which are focused on meeting quality requirements, proposing appropriate methods and tools to meet quality objectives, regularly checking them, and taking measures for continuous improvement [2]. From the consumer's point of view, we can define quality as the extent of the product's ability

M. Andrejiova · M. Pinosova (✉) · E. Lumnitzer
Technical University of Košice, Košice, Slovakia
e-mail: miriama.pinosova@tuke.sk

L. Knapčíková, D. Peraković (eds.), *6th EAI International Conference on Management of Manufacturing Systems*, EAI/Springer Innovations in Communication and Computing, https://doi.org/10.1007/978-3-030-96314-9_5

to perform the functions for which it was intended [4]. From the manufacturer's point of view, quality is associated with the technical level of the product in question under the prescribed conditions. The authors of the chapter use an integrated approach to the perception of the quality of the item from the customer's point of view and direct the relevant information into the product development process [5].

Failure modes and effects analysis (FMEA) is a useful tool for analyzing the production process. According to Stamatis [6], FMEA is a method used to define, identify, and/or eliminate known and/or potential faults, problems, and errors with the system, product, design, and/or services before they reach customers. It is used to prioritize three criteria: probability of failure, severity of failure, and detection of failure [7]. Vulpes T.C. et al. [8] state that FMEA is one of the most efficient and effective methods for preventing faults and other discrepancies. Its importance lies in assessing the possible failure of component functions during the manufacturing process, and their causes and effects in order to identify preventive measures that may result in higher reliability of the product.

Quality management methods also include statistical methods that use probability theory and processes from mathematical statistics to manage quality. According to Montgomery [1] and Vardeman and Jobe [9], statistical methods play an essential role in controlling and improving the quality of the production process, services, etc. These statistical process control (SPC) methods are used not only to measure and monitor process performance, but also to provide a basis for improving it. According to Mitreva et al. [10], by using SPC methods and cost optimization methodology it is possible to achieve defined quality and better productivity with the lowest operating costs. The seven basic tools of SPC include histograms, check sheet, Pareto chart, cause-and-effect diagram (Ishikawa diagram), defect concentration diagram, scatter diagram, and control chart [1]. These tools are easy and simple to use and can be applied to any process.

Filz et al. [11] used the FMEA method to improve scheduled maintenance. The FMEA method was also used to analyze the causes of deficiencies in the production of Covid-19 face masks using the 3D printing method [12]. In article [13], the authors apply the DMAIC (define, measure, analyze, improve, control) method, together with quality tools such as brainstorming, Ishikawa diagram, and Pareto chart, to analyze and reduce failure in some products' pneumatic brake system. Reliability, availability, and sustainability in a bag sector were analyzed by Tsarouhas [14]. He uses statistical quality management methods (descriptive statistics, Pareto chart, histogram, and others) to analyze production line failures. The optimization of the production chain, from the early stages of the design of production systems to the start of production of lithium-ion cells, was examined by Westermeier et al. [15]. They also used the Pareto chart in the evaluation and analysis of the production chain.

The authors of [16] used basic quality management tools (Ishikawa diagram, Pareto chart, and others) to identify the root causes of machine failures in production. In improving the injection molding process, the authors of [17] used quality management tools, namely statistical process control (SPC), Pareto chart, histogram, Ishikawa diagram, measurement system analysis, hypothesis test, and checklist. The statistical methods of quality control (Ishikawa diagram and Pareto chart) were also used in identifying the main factors influencing the process of pyrolysis of petroleum sludge, and also the causes of failures for which elimination is of essential importance [18]. Li et al. [19] use statistical process control and data mining technology for evaluations and improvements to the semiconductor element production process.

Optimization in the production process by introducing monitoring systems that allow the collection of production data, but also to report various failures or side effects, was addressed by the authors [20–22].

Quality control and improvement have become an important business strategy for many production organizations. Quality is a competitive advantage and therefore in every production process it is necessary to constantly implement quality management methods. The aim of this chapter is to analyze the production process using the selected quality management tools.

2 Material and Methods

2.1 Production Process

The company in which error analysis was carried out in the production process using quality management tools is a manufacturer of industrial electronics. It has been operating in Slovakia since 2000. The organization has an integrated management system and holds certificates that demonstrate compliance with the requirements of the ISO 9001:2015 (Quality management systems—Requirements), ISO 14001:2015 (Environmental management systems—Requirements with guidance for use), and ISO 27001:2013 (Information technology—Security techniques—Information security management systems—Requirements) standards. The basis of production is six assembly lines where the production of printed circuit boards (PCBs) takes place from their installation, testing, and final assembly to the delivery of the finished product to the customer.

One of the production lines of this company analyzed in this chapter is semiautomatic with manual production predominating. This is the finalization line on which PCB production is completed.

There are 11 operators working on the production line (see Fig. 1). They have workflows that describe individual work operations with the names and component labels for specific PCB models. Depending on the model currently being produced, operators choose a workflow and check the individual components (name, label, print). The first operator places an identification sticker on the PCB showing the model's name, date of manufacture, and 2D identification code. The tagged PCB then moves to the next operator who places the corresponding components on the PCB and moves it further along the line. In this way, the PCB board still goes

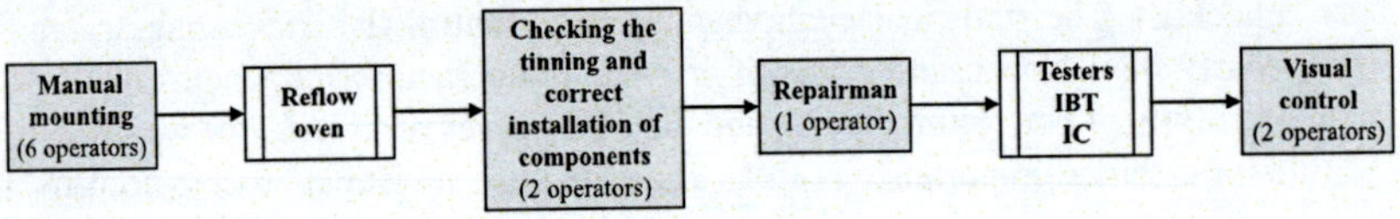

Fig. 1 PCB production process

through a manual process to the last operator, who checks the correct installation of all components and releases the board to the reflow oven.

The board goes further along the production line belt, passes the refrigeration equipment, and proceeds to another operator for bare-board testing which removes possible short circuits and open joints on the PCB. Another checks the correct installation of components and moves the PCB to the testers.

The first tester in turn is IBT. This is a test technique for testing installed printed circuit boards, which requires contact with each IC connector. PIN testers are also involved in IBT, which inspects the connectors of the installed components. The second is FT, which checks the correct function and prescribed PCB specifications. The board tested in this way continues to the final visual inspection. The PCB is visually inspected first from the hand side and then from the SMD side. The inspected board then goes into the box and is ready to pack.

If, during testing, one of the testers evaluates the PCB as defective, the IBT operator corrects the tester and inserts it for retesting. If the error cannot be removed, they tag the PCB with a sticker and take it to a specialized repair facility.

2.2 *Methods*

Three selected quality management tools (FMEA, Pareto chart, and Ishikawa diagram) are used to analyze the error rate of a particular production process.

FMEA is one of the basic tools applied within ISO 9000 standards. The aim of FMEA is to analyze potential errors in a particular system, over a selected period of the service life of the system, so that corrective measures can be taken to reduce the risk that the emergence of errors entails [23]. According to Stamatis [6], an FMEA system can be achieved using a series of steps that include conceptual design, detailed design and development, test, and evaluation.

Pareto analysis is one of the most effective quality management tools, which makes it possible to separate the root causes of a particular problem from the minor ones, thus showing the direction in which efforts to ensure quality assurance in the process need to be focused in order to remedy its shortcomings. Pareto analysis is based on the Pareto chart. The Pareto chart is named after the Italian economist Vilfredo Pareto (1848–1923), who considered that in certain economies a small proportion of the population held the majority of wealth [24]. The Pareto chart was first used in quality management by Juran (1904–2008), who stated that this principle can also be applied to errors, with 80% of the problems being caused by 20% of the defects [25]. A Pareto chart is a column chart of absolute, or absolute relative, occurrences of individual causes, in which are also plotted the points of the Lorentz curve representing the polygon of cumulative relative occurrences of each cause (in %). According to Montgomery et al. [1] the Pareto chart does not automatically identify the most important defects, but only the most frequent. Therefore, in addition to simple Pareto analysis, weighted Pareto analysis is often used, in which each type of error is assessed by the degree of severity of the error. Pareto analysis often uses the 80/20 or 75/25 rule, which means that it is recommended to include causes with a cumulative number from 0 to 80%, or from 0 to 75%, in a more detailed analysis of the causes.

Ishikawa (1915–1989) developed a simple tool to identify possible causes of a problem, known as a "cause-and-effect diagram" (or a fishbone diagram). Very often a version of the diagram with the main categories of causes is used: people, material, machines, methods, and environment [26].

3 Results and Discussion

The analysis and evaluation of the quality of the production process are divided into four parts:

- Analyzing the production process using the FMEA method
- Evaluating the occurrence of errors using Pareto analysis and Pareto chart
- Finding the main causes of the most common errors using the Ishikawa diagram
- Proposing measures to eliminate the most significant errors

3.1 FMEA of the Process

The aim of the analysis was to analyze the production process and to determine the occurrence of errors at the different stages of production. The main tool in the development of FMEA was brainstorming.

As part of process evaluation, the function and requirements of the process, possible defects (faults) of the product, possible consequence of the occurrence of errors, and severity of the errors were monitored. The research included analysis of the probability of the error occurring and its possible cause.

The result was the determination of a risk priority number (RPN), which is expressed as the product of three indices (severity, frequency, detection), where each index can receive a score in a certain interval. In our case, the score is on a scale of 1 to 5 (see Table 1). The resulting RPN value is between 1 and 25 points.

FMEA pointed to some critical points in the production process. Based on the RPN index, a risk level was established (low rate: RPN from 1 to 10 points;

Table 1 Rating for frequency, severity, and error detection

Rating	Occurrence	Severity	Detection
5	Very high	Very high	Very low
4	High	High	Low
3	Moderate	Moderate	Moderate
2	Low	Lower	High
1	Nearly impossible	None	Very high

Table 2 FMEA (RPN index ≥15 points)

Function/RPN	Possible cause/effect	Measures/RPN
Installation of ceramic capacitor (RPN = 20 points)	Connectors do not pass through the PCB hole/may cause the part to be thrown out or lifted during subsequent handling	Addition of control pin tester to IBT (RPN = 10 points)
Installation of AC inlet (RPN = 20 points)	Incorrect installation of the component/malfunction of the device	Installation of a camera to check pins, retrain employees (RPN = 10 points)
Installation of fuse (RPN = 20 points)	Unwetted area when installed/malfunctioning PCBs	Addition of pin tester (RPN = 10 points)
Installation of coil (RPN = 18 points)	Failure to follow the workflow/change of polarity on the device	Addition of polarity tester, retraining of employees (RPN = 12 points)

mean rate: RPN is from 11 to 14 points; high risk level: RPN ≥15b). The mean level of risk has been identified at production stages such as material preparation, glue application, capacitor, filter, and connector mounting. For risks exceeding a specified value of 15b, include 4 functions of the process (see Table 2).

FMEA pointed to some critical points in the production process. Measures to eliminate deficiencies were recommended for these production areas. The implementation of the measures has resulted in a significant reduction in the values of the RPN (see Table 2).

3.2 *Pareto Analysis*

In the analysis of errors in the production process, data from two time periods were used: 2019 (September, October) and 2020 (February, March), which were comparable in terms of production volume. Error analysis focused on the production line system for hand-installed components on the printed circuit board (PCB)—the final step of production. The errors observed, together with the degree of severity of the errors (weight of errors) obtained for the periods, are listed (see Table 3).

The degree of severity (determined with the assistance of experts from practice) is expressed using a natural number between 1 and 10, with 10 being the most serious error.

Table 3 Description of errors by weight for Pareto analysis

Designation	Description	Degree of severity (weight)
Err1	Missing component	7
Err2	Missing pin	10
Err3	Lifted component	8
Err4	Non-soldered area	9
Err5	Short circuit	8
Err6	Reversed polarity	9
Err7	Shifted component	7
Err8	Other errors	6

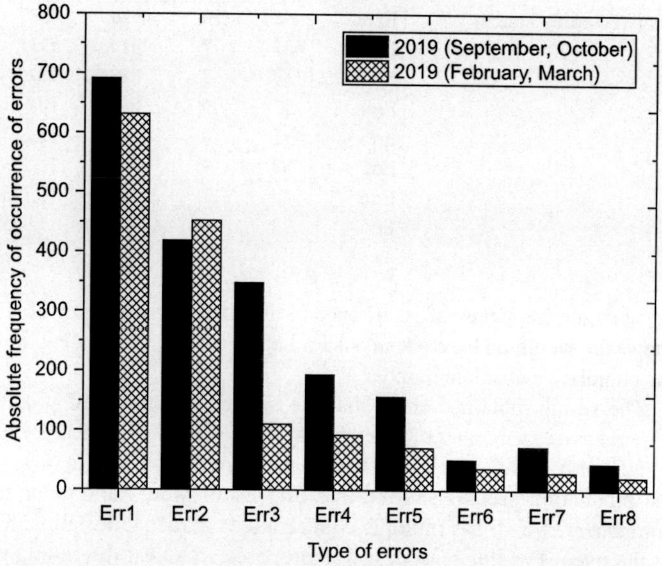

Fig. 2 Occurrence during the reference periods

Figure 2 is a graph showing the occurrence of errors in the manual installation of components during the reference periods. A total of 1978 deficiencies were identified in the 2 reporting months of 2019 (September and October). The most significant number is in the case of error Err1 (missing part), which accounted for almost 35% of the total number of noted errors. In 2020 (February and March), 1447 component planting defects were identified, representing a reduction of almost 27%. Again, the most significant number was error Err1 (missing part), which accounted for approximately 32% of the total number of noted errors in a given period.

For evaluation using Pareto analysis, we chose the 75/25 criterion, under which it is recommended to monitor in more detail errors for which the cumulative relative number for F_i reaches a value equal to 75%. We took into account not only the absolute frequency of occurrence of individual n_i errors, but also their degree of severity w_i (weight). We obtained the weighted number of wn_i errors as a product of

Table 4 Weighted Pareto analysis

Time period	Error	n_i	w_i	wn_i	N_i	F_i
2019 (September, October)	**Err1**	**691**	**7**	**4837**	**4837**	**0.312**
	Err2	**419**	**10**	**4190**	**9027**	**0.582**
	Err3	**348**	**8**	**2784**	**11,811**	**0.762**
	Err4	194	9	1256	13,067	0.843
	Err5	157	8	1200	14,267	0.920
	Err7	73	7	511	14,778	0.953
	Err6	51	9	459	15,237	0.983
	Err8	45	6	270	15,507	1.000
2020 (February, March)	**Err2**	**452**	**10**	**4520**	**4520**	**0.380**
	Err1	**631**	**7**	**4417**	**8937**	**0.751**
	Err3	111	8	888	9825	0.825
	Err4	93	9	837	10,662	0.896
	Err5	71	8	568	11,230	0.943
	Err7	37	9	333	11,563	0.971
	Err6	30	7	210	11,773	0.989
	Err8	22	6	132	11,905	1

the absolute frequency of occurrence of the error and its degree of severity. Table 4 shows the weighted Pareto analysis for both periods under review, which also shows the cumulative absolute number N_i.

The results obtained show that the situation in terms of product quality has improved over a number of months. Absolute error rates (except for Err2) decreased significantly. In the case of Err3 error, the decrease is by up to 68%. An analysis of the production process showed that Err3 errors were mainly due to the "cooler" component, which has multiple connectors, is harder to install, and can get raised in the oven. For this reason, it was proposed to adjust the setting of the oven by changing the shape of the weights used to load the cooler before it enters the oven. The situation also improved with the Err1 error (a decrease of almost 9% compared to the previous period). In the case of the Err2 error, the absolute number increased by almost 8% compared to the previous period. Therefore, for the Err2 error, it has been proposed to add an in-circuit board tester (IBT) control pin to check the length of the connectors. Adding this will extend the testing time but will improve the quality of the supplied board.

The Pareto chart shows that during the 2 months in 2019, three defects formed a vital group of causes of poor-quality products: Err1 (missing part), Err2 (missing pin), and Err3 (raised part). In total, these three errors accounted for up to 76% of all detected errors in the final stage of the PCB production process (see Fig. 3).

In 2020, the order of errors changed slightly. The vital group of causes consists of only two errors: Err2 (missing pin) and Err1 (missing part), which represent up to 75% of all errors in the production process during the given period (see Fig. 4).

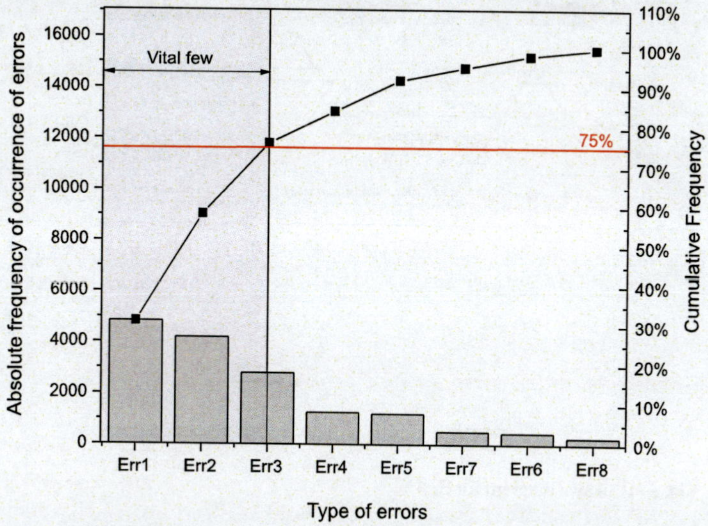

Fig. 3 Pareto chart (2019—September, October)

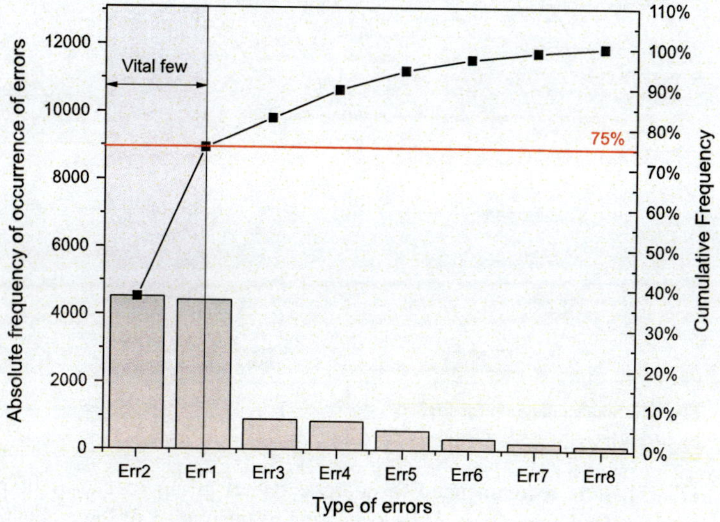

Fig. 4 Pareto chart (2020—February, March)

3.3 Ishikawa Diagram

The Ishikawa diagram or fishbone diagram is an appropriate tool for solving problems arising from the results of Pareto analysis. Under the chosen criterion

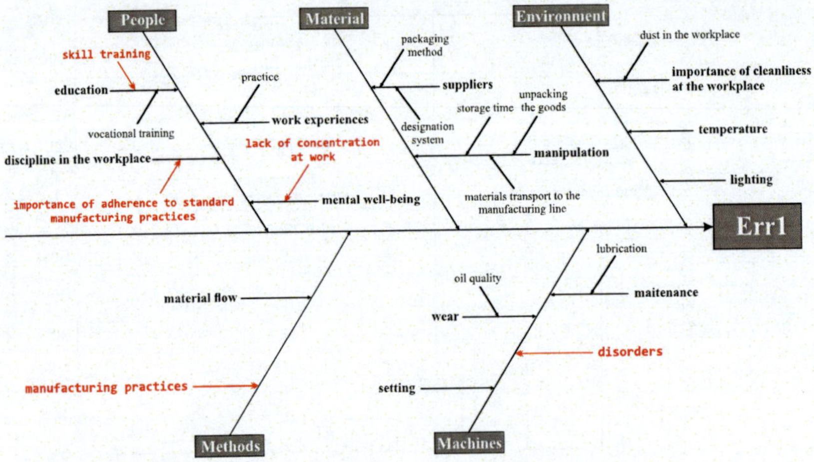

Fig. 5 Ishikawa diagram for Err1

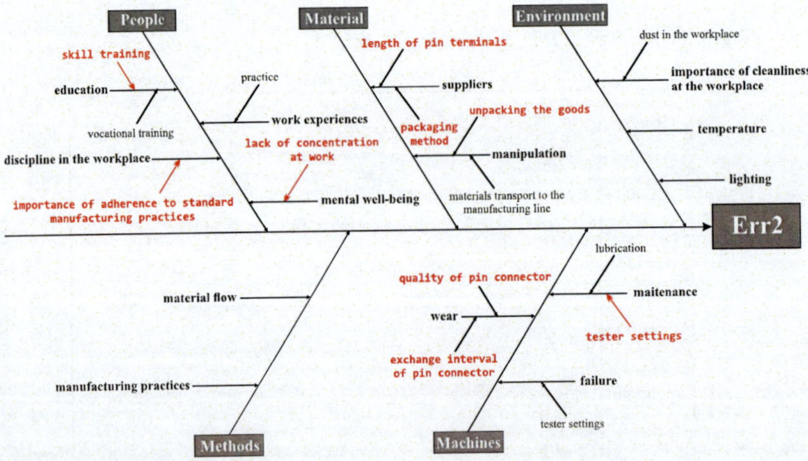

Fig. 6 Ishikawa diagram for Err2

(75/25), it is recommended to address a vital group of causes of poor-quality products in the next step of the evaluation of the quality of the production process. In the case of the monitored production process, these are errors Err2 (missing pin) and Err1 (missing part).

In each main category (people, material, machines, methods, environment), several causes were involved in the significant number of Err1 and Err2 errors, which are illustrated through the Ishikawa diagram (see Figs. 5 and 6). The basic sub-products for which measures have been proposed to eliminate Err1 or Err2 errors are indicated in red.

From the main category "Machines" Err1 error's significant causes are mainly due to belt and feeder faults. If the belt malfunctions, the part may be thrown from the printed circuit board before it is soldered. In the event of a failure of the feeder, the board is pushed out with the pressure piston "pusher" and if set up wrongly this can break off the part mounted from the surface mount device (SMD) side. The "Methods" category has an impact on the workflow that can be poorly or complicatedly written and misunderstood by the line operator. The cause most involved in the number of Err1 error is the "People" category, where compliance with working procedures, necessary concentration, and length of training are crucial.

The frequency of the Err2 error was mainly affected by the method of packaging and handling during unpacking and transport. Inappropriate packing of individual components, especially capacitors as the connectors on them are made of soft material, can easily lead to their being bent, resulting in more difficult mounting on PCBs. In case of incorrect unpacking, the connectors are bent, which also has an adverse effect on the quality of the PCB installation. In the case of the "Machines" category, the setting and maintenance of the tester, and also wear and tear of the control pin and the interval of its replacement and failure, are a significant cause. From the category "People" it is mainly work experience, working discipline, focus on the work performed, and adherence to the prescribed workflow.

3.4 *Measures to Eliminate the Most Significant Errors*

During the individual analyses of the error rate of the production process, categories of errors were specified for which measures had to be proposed and adopted to reduce or completely eliminate them. The proposed measures for the three most important main categories (People, Machines, Material) are summarized in Table 5.

Table 5 Summary of proposed measures

Category	Measure
People	Rewrite workflows into a more comprehensible and clearer form. Retrain operators and educate them about the internal rules of the organization. Inform operators about the number of errors caused by them and motivate them to work better. Set up training rooms for newly recruited operators where they will undergo 2 weeks of skill training before starting work
Material	Change the way in which the components are unpacked and supply capacitors to the operator on the line in their original packaging. Specify the length of the connectors. Add a pin control camera on the safety component
Machines	Shorten the interval for replacing contact pins in the tester. Consider changing the pin supplier for better quality. Always check the height of the control pin setting at the beginning of the shift, depending on the product type produced

4 Conclusions

For each production organization, quality is what the customer asks for and that is what every company is trying to provide. With quality tools, smooth, fast, and trouble-free production can be achieved by identifying a production problem area in a timely manner. Quality tools such as FMEA, Pareto analysis, and Ishikawa diagram may be used for this activity.

This chapter examines selected quality management tools and their use in the analysis of the error rate of the production of printed circuit boards. The individual procedures for analyzing a given production process identified errors for which it was necessary to design and take measures to reduce or eliminate them entirely.

Using the Risk Priority Index, FMEA has clearly highlighted the critical points in the production process for which the necessary measures have been proposed to prevent these errors. Pareto analysis identified the causes of key errors in the working environment, namely the Err1 and Err2 errors. An Ishikawa diagram is a very powerful tool for analyzing the causes of errors in the production process. For better diagnosis of the problem, for each major category of causes, subcategories were determined, supplemented by other minor causes with their share of the observed error. As part of the analysis, we addressed two of the most numerous errors Err1 and Err2. In both cases, it was confirmed to us that the human factor had a significant influence on the number of these errors. Therefore, human resources need to be given at least as much attention as technical equipment, finance, and other production capacities. Some of the proposed measures have already been implemented in the production process, but some require time to implement them, so they are currently being prepared. The aim of ongoing research is to verify the effects of the proposed measures on the quality of production.

Currently, a lot of attention is paid to the quality of any product. In addition to increasing requirements for increased and accelerated production, this is due to the growth of customer requirements for the quality of the final products. Deficiencies and defects will always appear in the production process. Defining significant errors and their causes enables your organization to focus on reducing or eliminating them. Quality management methods play an important role in the analysis and evaluation of errors.

Acknowledgments This chapter was prepared with the support from the project titled KEGA No. 011TUKE-4/2021: Implementation of current scientific-research, technical and methodological solutions in the environmental engineering field into the educational process at universities.

References

1. D.C. Montgomery, *Statistical Quality Control*, 6th edn. (Wiley Global Education, New York, 2008)
2. ISO 9000:2000 Quality management systems – Fundamentals and vocabulary. Homepage, https://www.iso.org/standard/29280.html. Accessed 9 May 2021

3. R. Kumar Sharma, R. Gopal Sharma, Integrating Six Sigma culture and TPM framework to improve manufacturing performance in SMEs. Qual. Reliab. Eng. Int. **30**(5), 745–765 (2014)
4. J. Berk, S. Berk, *Quality Management for the Technology Sector* (Elsevier, Woburn, 2000)
5. B. Falk, B. Quattelbaum, R. Schmitt, *Product Quality from the Customers' Perspective–Systematic Elicitation and Deployment of Perceived Quality Information*, in Proceedings of the Proceedings of the 6th CIRP-Sponsored International Conference on Digital Enterprise Technology (Springer, Hong Kong, 2010), pp. 211–222
6. D.H. Stamatis, *Failure Mode and Effect Analysis: FMEA from Theory to Execution*, 2nd edn. (Quality Press, Wisconsin, 2003)
7. H.-C. Liu, *FMEA Using Uncertainty Theories and MCDM Methods* (Springer, Singapore, 2016)
8. T.C. Vulpes, C.G. Opran, *Research on Optimizing of the Failure Mode and Effects Analysis Using Advanced Expert System*, ed. by G. Lamanna, C.-G. Opran. 4th international Conference POLCOM 2020 – Progress on Polymers and Composites Products and Manufacturing Technologies, POLCOM, vol 396 (Wiley-VCH GmbH, Romania, 2021), pp. 1–4
9. S.B. Vardeman, J.M. Jobe, *Statistical Methods for Quality Assurance: Basic, Measurement, Control, Capability and Improvement*, 2nd edn. (Springer, New York, 2016)
10. E. Mitreva, S. Gjurevska, N. Taskov, H. Gjorshevski, Application of methods and techniques for free production in a company for manufacture of bread and bakery products. Qual. Access. Success. **20**(171), 141–145 (2019)
11. M.-A. Filz, J.E.B. Langner, C. Herrmann, S. Thiede, Data-driven failure mode and effect analysis (FMEA) to enhance maintenance planning. Comput. Ind. **129**(2021), 1–17 (2021)
12. D.D. Rochman, R.R. Margana, A. Anwar, R. Ferdian, Application failure mode effect analysis on 3D printing Covid-19 face mask. Turkish J. Comput. Math. Educ. **12**(8), 1723–1728 (2021)
13. T.S. Duarte, S.B.P. Maia, L. de Césares Martins, B. Mihailoff, L.C. Batista, E.B. Salvador, B.P. de Campos, *Application of Quality Tools to Reduce Pneumatic Leaks in the Commercial Vehicle Brake System*, in 2020 SAE Brasil Congress & Exhibition (SAE Technical Paper, Sao Paulo, 2021), pp. 1–9
14. P. Tsarouhas, Reliability, availability and maintainability analysis of a bag production industry based on the six sigma DMAIC approach. Int. J. Lean. Six Sigma **12**(2), 237–263 (2020)
15. M. Westermeier, G. Reinhart, M. Steber, *Complexity Management for the Start-Up in Lithium-Ion Cell Production*, ed. by R.H. Schmitt. 2nd International Conference on Ramp-Up Management 2014, ICRM, vol 20 (Elsevier, Germany, 2014), pp. 13–19
16. M.A. Amrani, M. Alhomdi, A.M. Ghaleb, M. Al-Qubati, M. Shameeri, Implementing an integrated maintenance management system for monitoring production lines: a case study for biscuit industry. J. Qual. Maint. Eng. (2020)
17. A. Maged, S. Haridy, S. Kaytbay, N. Bhuiyan, Continuous improvement of injection moulding using Six Sigma: case study. Int. J. Ind. Syst. Eng. **32**(2), 243–266 (2019)
18. O.A. Kolenchukov, E.A. Petrovsky, *Analysis of the Causes of Failures of Pyrolysis Units*, in International Scientific Conference on Applied Physics, Information Technologies and Engineering 2019, APITECH, vol 1399 (IOP Publishing, Russian Federation, 2019), pp. 1–5
19. B. Li, R.-S. Chen, C.-Y. Liu, Using intelligent technology and real-time feedback algorithm to improve manufacturing process in IoT semiconductor industry. J. Supercomput. **77**(5), 4639–4658 (2021)
20. L. Knapčíková, A. Behúnová, M. Behún, Using a discrete event simulation as an effective method applied in the production of recycled material. Adv. Prod. Eng. Manag. **15**(4), 431–440 (2020)
21. L. Knapčíková, A. Behúnová, M. Behún, Workplace optimization in the manufacturing enterprise supported by digitalization software. TEM J. **8**(4), 1259–1263 (2019)
22. A. Behúnová, M. Behún, L. Knapčíková, TEM J. **7**(4), 849–8309 (2018)
23. J. Sinay, H. Pačaiová, J. Grenčík, A. Golianová, Š. Markulik, A. Mokrišová, A. Nagyová, B. Treštíková, *Nástroje Zlepšovania Kvality (in Slovak)*, 1st edn. (ManaCon, Prešov, 2007)

24. D. Parmenter, *Pareto's 80/20 Rules for Corporate Accountants*, 1st edn. (Wiley, New York, 2007)
25. M. Best, D. Neuhauser, Joseph Juran: overcoming resistance to organisational change. BMJ Qual. Saf. **15**(5), 380–382 (2006)
26. E. Jones, *Quality Management for Organizations Using Lean Six Sigma Techniques*, 1st edn. (CRP Press, London, 2019)

Part II
Sustainable Communications and Information and Communications Technology

A Taxonomy of Cyber Attacks in Smart Manufacturing Systems

Bethanie Williams, Marena Soulet, and Ambareen Siraj

1 Introduction

According to the National Institute of Standards and Technology (NIST), smart manufacturing is "fully integrated, collaborative manufacturing systems that respond in real time to meet changing demands and conditions in the factory, in the supply network, and in customer needs" [1]. Manufacturers and companies see smart manufacturing as a profitable market opportunity in this modern era. It is also referred to as Industry 4.0, which is being globally adopted throughout the manufacturing sector. Industry 4.0 can be considered as the next-generation manufacturing model that integrates advanced information and communication technologies (ICT) to further improve the overall process, efficiency, and profits in the manufacturing systems.

Although the shift in smart manufacturing is gaining competitive advantages in the global market, the increase in the frequency and sophistication of cyberattacks within the manufacturing sector poses a significant threat to companies and organizations everywhere making security challenges in smart manufacturing a global concern. According to Michael Taylor, "Manufacturing was one of the top five most targeted industries by adversaries in 2016, 2017, and 2018" [2]. However, due to lack of awareness, resources, and/or enforcement, many manufacturers lack appropriate security measures to counter against cyber threats and attacks.

Over the past decade, there have been several cyberattacks that have targeted industrial control systems (ICS). One of the most recent attacks occurred in a water treatment plant in Oldsmar, Florida, in which a hacker breached computer networks and increased the amount of sodium hydroxide in the water supply with potential

B. Williams (✉) · M. Soulet · A. Siraj
Tennessee Technological University, Cookeville, TN, USA
e-mail: bdwilliams44@tntech.edu; mbsoulet42@tntech.edu; aisraj@tntech.edu

L. Knapčíková, D. Peraković (eds.), *6th EAI International Conference on Management of Manufacturing Systems*, EAI/Springer Innovations in Communication and Computing, https://doi.org/10.1007/978-3-030-96314-9_6

adverse effect on public health [3]. The most notable attack in ICS was Stuxnet, which was a malware that "infected the software of at least 14 different industrial sites in Iran in 2010 and caused physical damages to a fifth of Iran's nuclear infrastructure" [4]. Stuxnet could have been avoided with appropriate security protections in place for the targeted systems with technology solutions such as encryption, access control, and key management. Another example is a well-known incident that occurred in the Ukraine power grid in 2015, in which an attack caused over 100,000 people to be without power and which could have been prevented if there were proper two-factor authentication in place. All these incidents that resulted in physical damages, monetary losses, and/or compromised production may have been averted if standard security protocols and procedures were enforced.

Cybersecurity needs in smart manufacturing must play a vital role in protecting its human operators and users, environment, products, production processes, and trade secrets such as data or process designs from bad actor interventions. Furthermore, to aid manufacturers to be successful in protection against cyber threats and attacks, the NIST has developed a cybersecurity framework particularly for the manufacturing industry [5]. When followed, the standard can help manufacturers become more aware of potential security challenges, reduce cybersecurity risks, detect attacks as they occur, and learn how to recover from cyberattacks. The framework provides manufacturers with awareness and protection tools to meet the security needs of their manufacturing system. It applies to manufacturing systems of all sizes, from large corporations to small businesses, with fundamental security protocols and procedures, ensuring protection against cyber threats and attacks.

In this chapter, with the NIST cybersecurity framework for manufacturing standards in mind, we propose a taxonomy of the most common types of attacks that have been observed and reported in smart manufacturing today. This will help both manufacturers and researchers gain a deeper understanding of smart manufacturing attacks and how to secure systems against such attacks as per cybersecurity framework NIST standard. The remainder of the survey chapter is organized as follows: In Sect. 2, the NIST Cybersecurity Framework Manufacturing Profile is discussed. A perspective based upon the NIST's manufacturing objectives and categorization process along with the sources we investigated and analyzed in our research is discussed in Sect. 3. In Sect. 4, we present a taxonomy of reported cyberattacks based on the NIST's manufacturing framework's objectives and classifications of security levels. Through the framework, impacts of manufacturing cyberattacks are discussed, analyzed, and labeled. Finally, conclusion and future research areas are presented in Sect. 5.

2 Standards of the NIST Manufacturing Architecture

2.1 NIST's Manufacturing Objectives

The NIST Cybersecurity Framework Manufacturing Profile (NIST-CFMP) is essential to developing a more secure architecture for manufacturers worldwide. It is highly valuable to manufacturers and researchers who are using technology to enhance the security of manufacturing processes. The NIST-CFMP mainly focuses "on the desired cybersecurity outcomes and provides an approach to the desired state of cybersecurity posture of the manufacturing system" [5]. The primary reason the NIST constructed this framework is to help support the missions of the manufacturing industry as well as develop a guide for the best strategies and techniques to secure manufacturing systems. If manufacturing companies and organizations follow the guidelines provided by the NIST, it will most likely contribute to reduction of cybersecurity threats, risks, and attacks to their organizations in great extents. This framework acts as a roadmap for all manufacturers who are interested in using smart manufacturing. Moreover, it is also used to help manufacturers improve their cybersecurity standards or protocols currently implemented in their systems.

As technology is constantly evolving, manufacturing systems will only become more enhanced with technology, causing an increase in security concerns with misuse/abuse of technology. Therefore, security should be a primary goal of future manufacturing systems. It is impossible to detect every possible attack from happening; however, by implementing the NIST's standard guidelines, the likelihood of successful attacks and damages caused by the attack can be drastically reduced. The NIST-CFMP provides manufacturers with a "standardized approach to preparing the cybersecurity plan for ongoing assurance of the manufacturing system's security" [5]. Manufacturing systems should utilize the framework to help them identify, protect, detect, respond, and recover from possible security issues. The future of secure smart manufacturing systems relies on successful and widespread adoption of the NIST-CFMP.

Through the development of cybersecurity manufacturing profiles, the NIST can identify a set of primary objectives for secure smart manufacturing industries. These objectives supply imperative factors for recognizing and handling relevant cybersecurity practices set in place to reduce risk. The five most common objectives identified for the manufacturing sector are maintaining human safety, environmental safety, quality of product, production goals, and trade secrets [5]. Each of these objectives provides manufacturers insight into how cyber threats can cause harm and affect their manufacturing systems. The listed objectives are also accompanied by suggestions of what and how to protect based upon the objective. The objectives, along with an explanation of their importance in the smart manufacturing sector, are summarized in Table 1.

These objectives play a vital role in providing the manufacturing industry with insights on what requires protection and the proper tools to protect such against

Table 1 NIST-CFMP [5] objectives summarized

Objectives	Description	Importance
Maintaining human safety	Being aware of potential concerns that may cause an individual to be impacted by serious harm and effectively mitigating cybersecurity risks to ensure human safety at all times	Ensures that no harms such as injury or death happen to employees as a result of a cyberattack
Maintaining environmental safety	Being aware of potential concerns that may cause the environment to be negatively affected and effectively mitigating cybersecurity risks to ensure environmental protection from harmful incidents caused accidentally or deliberately	Ensures conservation of natural resources and environment
Maintaining quality of product	Being aware of potential concerns that may cause the quality of product to be defected and effectively mitigating cybersecurity risks to ensure quality of the product by preventing product defects, improper assembly, falsifying data, product designs, etc.	Ensures that customer trust and satisfaction are maintained and costs of reproducing products are minimized or avoided
Maintaining production goals	Being aware of potential concerns that may hinder production goals and effectively mitigating cybersecurity risks to ensure efficiency, performance, production, flow, and low costs within the production process	Ensures maintenance of strict deadlines, good reputation, and monetary benefits
Maintaining trade secrets	Being aware of potential concerns that may cause trade secrets to be exploited and effectively mitigating cybersecurity risks to ensure protection of an organization's secrets such as data, designs, other intellectual property, or any other confidential information	Ensures maintenance of competitive edge, sensitive materials, and secrecy

cyber threats, risks, and attacks. If one of these objectives is not properly addressed and protected, there could be dire consequences. For example, cybersecurity threats can jeopardize human life if safety protocol is compromised by carrying out an attack. Security breaches can also cause harm to the physical environment if machines are damaged. Products having defects, production being stopped, and data being compromised are all results of a standard objective not being met.

2.2 NIST's Manufacturing Objectives

The NIST's cybersecurity framework for manufacturing developed a distinctive process for categorizing cyberattacks based upon their potential risks and/or impacts. It can be broken down into three different types of security levels: low, moderate, or high. These categories help identify the security measure's performance, relevance, and effectiveness for assessing the level of risk. The security levels allow attacks to be categorized by the significance of the harm that they may cause to the manufacturer, which also allows manufacturers to prioritize the protection needs for their assets. Table 2 is a brief view of each security level and its significance to the manufacturing industry.

The main goal of this categorization process is to help organizations identify the security level to which a cyberattack belongs when a manufacturing system or process is jeopardized. To further analyze these three categories of security, the NIST-CFMP uses the CIA triad (confidentiality, integrity, and availability) to denote the type of impact a threat may potentially cause for the manufacturing system. This categorization process is essential in helping manufacturers decide what security measures need to be implemented to ensure CIA protection goals. For example, an organization trying to maintain a manufacturing system with its vital operations, data, and assets will need to take the appropriate security measures to ensure protection and mitigate potential risks to all of the above. Thus, one can see how the categorization process of labeling reported cyberattacks according to the NIST-CFMP can be useful for industries in the manufacturing sector.

Table 2 NIST-CFMP security levels' explanation

The potential impact is ...	If losses of the CIA triad are expected to have a ____________ on manufacturing operations, products, brand images, finances, personnel, environment, or general public
Low	Limited adverse effect
Moderate	Serious adverse effect
High	Severe or catastrophic effect

3 An Attack Taxonomy Based on the NIST Manufacturing Objectives and Security Categorization Processes

In the previous section, we provided an overview about the NIST manufacturing objectives and their importance along with a discussion of the NIST's categorization process with the three primary security levels. Now, we share our own perspectives and thoughts on the subject.

The primary focus of this study is to create a taxonomy of smart manufacturing cyberattacks by using the NIST's framework. We have mapped smart manufacturing cyberattacks to both the manufacturing objectives and the categorization process based on security levels. The proposed taxonomy is used to categorize smart manufacturing cyberattacks based upon the priorities of the manufacturing objectives, security challenges, and potential impacts on the manufacturing systems.

A cyberattack can cause a manufacturing system to fail in providing one or more of its objectives identified in the NIST-CFMP such as protection of human safety, trade secrets, or production process. Many of the cyberattacks mentioned in this study have been identified, discussed, and investigated by various public entities, news outlets, government agencies, and international organizations. Some of these organizations include the US Homeland Security, Federal Bureau of Investigation (FBI), National Security and Defense Council of Ukraine, the Australian Police Department, and others. This survey was conducted through an analysis of publicly available technical publications to establish a baseline for cyberattacks in smart manufacturing. We first manually scanned through all the major databases and publications which include conference proceedings, journals, and newsletters on the subject. A few notable databases we explored include ACM Digital Library, Applied Science & Technology database, ArXiv.org (Cornell University Library), Computer Science Database, DOE CODE, DOE PAGES, IEEE/IET Electronic Library (IEL), Science.gov, and Telecommunications Database. To ensure that this survey covers the latest reported incidents, we further selectively searched through Google Scholar. This resulted in a total of 48 papers, published from 2010 until 2021, that discussed reported attacks within the manufacturing sector over the last 20 years.

Through our research, we discovered that many of the attacks targeted at manufacturing organizations aim to compromise their cyber systems with the intention to target and cause harm to the physical environment. Thus, most of the attacks mentioned within the taxonomy did or could have caused physical damages using the cyber systems. If the attack executed did not ultimately cause significant physical damages, it is because the level of impact was low or because the attack was related to gaining digital assets or trade secrets. In some cases, there could have been serious physical damages if the attacks had not been detected or responded to in time.

Another pattern we recognized was the manufacturing sector's vulnerabilities found in industrial control systems (ICS) and information technology (IT) systems. Attackers seem to place a high emphasis on targeting ICS through supervisory control and data acquisition (SCADA), programmable logic controllers (PLCs), distributed control systems (DCS), human-machine interfaces (HMI), and other controls that are used to operate the industrial processes. By attacking the components of ICS, hackers can cause extreme chaos and damages such as operational shutdowns, safety hazards, theft of intellectual property, and destruction of equipment. The other highly targeted area that attackers focus on is IT. An organization's operations can be adversely affected if an attacker gains access to a manufacturer's entire network, causing harm to its hardware, software, and data repositories residing in the network. Common threats in IT can include various malicious software such as worm, trojans, ransomware, spyware, and other viruses.

Being able to recognize these adversarial activities and being aware of the nature of these attacks allow researchers and manufacturers to prepare themselves against these types of attacks. Having the capabilities to identify highly targeted areas in manufacturing will enable organizations to prioritize their security needs. Increasing security measures and having proper security protocols in place for ICS and IT systems will play an important role in ensuring protection within the manufacturing process.

Over the last few months, our investigative study resulted in the proposed taxonomy, which identifies, defines, and classifies cyberattacks and can be used by both researchers and manufacturers to determine which objective(s) is/are at risk as a result of such attacks. The NIST's categorization will additionally allow evaluation of the extent of such risks and impacts upon the systems. Furthermore, a list of possible attacks that have reportedly occurred will aid manufacturers to recognize similar attacks and better understand the nature and magnitude of potential damage to the overall production processes and systems. Thus, this would enable manufacturers to develop a proactive approach to mitigate these attacks, resulting in increased security in smart manufacturing systems.

4 Taxonomy of the NIST Manufacturing Objectives with Categorization of Security Levels

Following is a description of the proposed taxonomy that is inspired by the NIST's Cybersecurity Framework Manufacturing Profile. The uniqueness of the taxonomy lies in using the NIST's manufacturing profile to classify possible cyberattacks applicable in the entire manufacturing sector. An extensive literature review of cyber-physical security incidents in manufacturing and their impact has been studied in developing this taxonomy. By combining the NIST's manufacturing objectives with the categorization process, we have cataloged various cyberattacks discovered in this research study. Our hope is that this taxonomy brings awareness about cyberattacks that have occurred in the past to both researchers and manufacturers in the industry, accompanied by guidance to protect against similar attacks in the future. The taxonomy uses the following attributes:

Incident	→	A cyber event capable of jeopardizing the CIA goals of information or physical systems
Attack mechanism	→	A method or strategy used to target and compromise a system
Level of impact	→	Extent of potential damage caused by a cyberattack
CIA violation	→	A violation of one or more of the CIA goals: confidentiality, integrity, and availability
Type of threat	→	A potential agent/action that causes deception (falsification of data/source/process), disclosure (leakage of sensitive information), disruption (interruption of data/process), and usurpation (adversarial control of data/process)
Actors	→	Groups or individuals who purposely aim to exploit a system's vulnerabilities to gain unauthorized access to information and/or cause harm to a system

4.1 *First Dimension: Human Safety (Table 3)*

Low-Impact Vector: Davis-Besse Nuclear Power Station Worm Infection—In 2003, a computer worm known as the Slammer worm exploited a vulnerability in Microsoft SQL Server, resulting in over 75,000 servers becoming infected within minutes. The worm hid within system memory and searched for other hosts to contaminate. According to Kesler, "It searched for new hosts by scanning random IP addresses. This generated a huge volume of spurious traffic, consuming bandwidth and clogging networks" [6]. It proceeded to infect computer systems at the nuclear power plant near Oak Harbor, Ohio. The worm found its way in through a secondary pathway into the control network over a consultant's network, avoiding all the access control policies enforced by the organization's firewall. The infection overloaded the traffic on the site network. As a result, the Safety Parameter Display System (SPDS), which displays highly pertinent data regarding the reactor core, was unavailable for several hours. Although the Slammer worm carried no malicious payload, it still caused substantial disruption, which is why it can be classified as a low-impact vector. While the worm was able to successfully block sensors from displaying information digitally to control systems, there were two things that greatly reduced the risk of the attack: it did not have an effect on the analog readouts on the equipment itself, which meant that personnel could still obtain correct data from the physical sensors. The second was that the plant was not active at the time the attack occurred, so there was no big loss or human endangerment.

Moderate-Impact Vector: Jeep Cherokee Ignition Switch—In 2015, it was announced that hackers had taken control of a Jeep Cherokee through its Internet-accessible entertainment system. The vulnerability impacted over a million vehicles purchased in the United States, which had been equipped with the company's Bluetooth system. Through a hacking technique classified as a zero-day exploit, the hackers were able to execute code, which allowed them to wirelessly break into the vehicle's system and grant them access to control vital functions such as its steering, brakes, and transmission [7]. This can be classified as a moderate-impact vector, as

Table 3 Cyber-physical attack events and their impact on human safety

Incident	Attack mechanism	Level of impact	CIA violation	Type of threat	Actors
Car Shark	Malicious Software	High	IntegrityAvailability	UsurpationDisruption	Domestic (USA)
Davis-Besse Nuclear Power Station	Worm	Low	Availability	Disruption	Foreign (Czech Republic)
Jeep Cherokee Ignition Switch	Zero-day exploit	Moderate	ConfidentialityIntegrity	Usurpation	Domestic (USA)
Saudi Arabian Petro-chemical Plant	Malware	High	Integrity	UsurpationDisruption	Foreign (Russia)

the insufficiency of necessary security measures in place to protect drivers against hackers can endanger human lives and have potential serious consequences.

High-Impact Vector: CarShark—In 2010, two teams of researchers from the University of Washington and the University of San Diego designed a program named CarShark that allowed them to hack into a car's computer system. Two tests were conducted. The first test, which was done wirelessly, "was done by taking advantage of the sensors inside each tire that broadcast a brief radio signal every 60 to 90 seconds" [8]. The signal was able to communicate the pressure of each tire to the vehicle's computer system and once intercepted, the researchers sent erroneous low air pressure warnings to the car's dashboard screen and eventually crashed the internal computer. In the second test, researchers were able to physically hack into a vehicle's computers by directly connecting to them. They were able to again access internal vehicular networks that oversee controlling the engine, brakes, and door locks. This is alarming, as humans could potentially be harmed if malicious hackers learn to exploit or invade a vehicle's control systems from a distance.

Saudi Arabian Petrochemical Plant—In December 2017, it was announced that an oil and gas petrochemical center in Saudi Arabia had experienced an emergency shutdown as the result of a malware attack. The malware, termed Triton (also known as Trisis/HatMan), exceeded other industrial cyberattacks by targeting safety-instrumented systems into running unauthorized programs on the Safety Instrumented System (SIS). As Walker states, "by infecting the SIS controller, it's believed the malware could reprogram the parameters, forcing unsafe machinery to continue to operate, potentially causing physical damage to the equipment or even injury to human operators" [9]. Considered as a high-impact vector, this incident resulted in millions of dollars in lost production, and posed a threat to human life, as dozens of people could have been killed.

4.2 *Second Dimension: Environmental Safety (Table 4)*

Low-Impact Vector: Rye Brook, New York Dam—In 2013, a group of Iranians targeted a small dam in New York. This dam is only about 20 ft tall and is used for flood control [10]. The hackers were not able to cause any physical damages to any aspect of the dam, town, or environment. However, they were able to access information about the "dam's operations including its water level, temperature, and status of the sluice gate" [11]. The Iranians who planned this attack were eventually charged for committing the cybercrime [12]. This is considered a low-impact event regarding environmental safety because the Iranians were able to gain access to the dam's system but did not cause any physical damages. However, imagine if this attack had been on the Hoover Dam, this would have caused large-scale flooding in urban and rural areas throughout Nevada.

Florida Water Treatment Plant—In February 2021, a breach occurred at a water treatment facility located in Oldsmar, Florida. The hacker was able to gain remote access to the water treatment plant's SCADA controls through TeamViewer, a

Table 4 Cyber-physical attack events and their impact on environmental safety

Incident	Attack mechanism	Level of impact	CIA violation	Type of threat	Actors
Florida Water Treatment Plant	Unauthorized Remote Access	Low	Integrity	Usurpation	Unknown
Australia Wastewater Treatment Plant	Insider Threat	High	Integrity Availability	Disruption Usurpation	Domestic (Australia)
BTC Turkey Pipeline Explosion	Malicious Software	Moderate	Integrity Availability	Disruption Usurpation	Russia
US Turbine Control System	Malware	Moderate	Confidentiality Availability	Disruption Disclosure	Unknown
New York Dam	Unauthorized Remote Access	Low	Confidentiality	Disclosure	Iran
Ukraine Power Grid	Malware	High	Availability	Disruption	Russia

remote-access software, and changed the level of sodium hydroxide in the water from 100 parts per million to 11,100 parts per million, which is potentially harmful to humans if ingested in large amounts [3]. Fortunately, the plant operator monitoring the water plant noticed the breaches occurring and immediately reduced the levels back to safe measures before any harm was done. Investigators were able to discover that not only was the Oldsmar water treatment system using the 32-bit version of the Windows 7 operating system, but also all computers using the remote-access software shared the same password credentials and appeared to be connected to the Internet without any type of protection mechanism installed [13].

Moderate-Impact Vector: Baku–Tbilisi–Ceyhan (BTC) Turkey Pipeline Explosion—In 2008, a cyberattack caused a massive explosion near Refahiye, Turkey. It is believed that the attackers exploited the camera system which led to gaining access to the pipeline's network. This allowed the attackers to target "industrial computers (most likely remote terminal units (RTUs) or programmable logic controllers (PLCs)) at valve stations to change pressure and misreport values back to the control room" [14]. The hackers who planned the attack were very knowledgeable. They were able to turn off alarms and increase the pressure of crude oil in the pipeline to cause a major explosion that resulted in the spilling of over 30,000 barrels of oil [15]. The explosion resulted in massive damages to the pipeline, town, and environment.

US Turbine Control System—In 2012, a computer virus infected an unidentified US power plant's turbine control system. The virus was infected by a technician through a USB drive, which hindered operations at the power plant for 3 weeks and caused substantial economic loss [16]. However, the employee did not know that the USB had a virus on it. The Department of Homeland Security discovered the malware in the USB to be a variant of the Mariposa virus, which can be used to steal classified information from compromised computers such as data, account details, and log-in credentials [17]. As a result, the attack affected about 10 computers on the system's network and shut down the plant for 3 weeks [18]. This cyberattack can be classified as a moderate-impact incident due to the plant's activeness in recovering from the incident. It did not restrict the plant from producing electricity.

High-Impact Vector: Wastewater Treatment Plant in Maroochy Shire, Australia—In 2000, the Maroochy Shire Council in Australia started showing signs of a cyberattack taking place within its wastewater system. The system experienced problems that included communication failures between pumping stations and the main computer system; pumps were not running properly, and alarms were not reporting to the system engineers [20]. It took weeks for someone to finally notice and conclude that someone was intentionally causing these problems. Eventually, the police and a group of private investigators were able to track down the hacker who was interfering with the system. This attack resulted in a spill of 264,000 gallons of sewage to local areas such as parks, rivers, and even a hotel [20]. The impact of this attack caused detrimental harm to the environment.

Ukraine Power Grid—In 2015, a cyberattack on the Ukraine power grid affected close to a quarter-million Ukrainian citizens by cutting off their electricity [19]. Hackers brought down the power supply of the Ukrainian homes by using different versions of malware. The malware used in the attack was called BlackEnergy, and the attack pointed to a group of Russian hackers known as the Sandworm team. This Russian team had previously infected power suppliers in both the United States and Europe. It said that the "virus attack" had been accompanied by "floods" of calls to the companies' technical support numbers [21]. This cyberattack is one of the most significant incidents that have occurred in cyber history because it was "the first publicly acknowledged incident to result in power outage" [22]. It is important to remember how a large number of Ukrainian people were impacted without power for hours in their homes. This cyberattack sent a message to all organizations in the energy sector to maintain proper protection from cyber threats.

4.3 Third Dimension: Quality of Product (Table 5)

Low-Impact Vector: Case Study Virginia Tech—In 2014 and 2017, two case studies were conducted at Virginia Tech to demonstrate in a controlled environment if a cyber-physical attack was capable of successfully enabling a production system to produce manufactured products that were notably different from their original designs. In the first case study, tool path files were altered for a machining process [23] resulting in considerably flawed and incorrect part dimensions. It is noteworthy to mention that the attack was planned and executed successfully without alerting any of the personnel inspecting the product. In the second case study, design files

Table 5 Cyber-physical attack events and their impact on quality of product

Incident	Attack mechanism	Level of impact	CIA violation	Type of threat	Actors
Virginia Tech Case Studies	Code Modification	Low	Integrity	Disruption Deception	Domestic (USA)
Dyn Inc.	Denial of Service (DoS)	Moderate	Integrity Availability	Disruption Deception	Domestic (USA)
Kemuri Water Plant	Malware	High	Confidentiality Integrity	Disruption Disclosure	Foreign Hacktivist Group
WannaCry Virus	Ransomware	High	Integrity Availability	Disruption Disclosure	North Korea

for an additive manufacturing process [24] were altered to produce an internal void within a product, resulting in the design intent being altered and an eventual failure in the product. It also went undetected by manufacturing inspection systems, which is a concern due to the fact that an attack similar to the one carried out in these case studies could not only negatively affect a product's design intent, performance, and quality, but also pose a safety risk to consumers exposed to faulty products.

Moderate-Impact Vector: Dyn Inc.—In 2016, cyberattackers performed a series of denial of service (DoS) attacks on Dyn Inc. This attack was able to shut down a majority of America's Internet, and was caused by a form of malware called the Mirai botnet [25]. The hackers who performed this sophisticated attack were able to take control of various IoT devices including CCTV cameras manufactured by a Chinese company called Hangzhou Xiongmai Technology. The botnets of the attack were able to generate fake network traffic to cause Dyn Inc.'s network to crash [26]. The attack resulted in the Chinese manufacturer recalling 4.3 million devices sold to the United States because of a security problem with the devices' default passwords [27]. It is believed that the attack Dyn Inc. experienced was caused by these devices through the use of the botnets. The impact of the attack is considered moderate since the company was able to detect the attack and recall products. However, this cyber incident did cause the Internet to shut down for a day in multiple areas throughout the United States.

High-Impact Vector: WannaCry Virus—In May 2017, a ransomware worm spread swiftly across several computer networks. The massive global attack "infected more than 300,000 computers in 150 countries around the world, including at the UK National Health Service, the Russian Ministry of Interior, the Deutsche Bahn railway, and global companies such as Nissan, Renault, and FedEx" [28]. The ransomware took advantage of a security vulnerability in Microsoft's operating system, with intelligence agencies tracing the virus to a hacking group linked to North Korea. The crypto worm, after successfully infecting a Windows computer, would encrypt files on the machine's hard drive, making them inaccessible for the user. It would then pressure the victim into paying a ransom payment in order to decrypt them. To date, WannaCry is one of the largest and most catastrophic ransomware attacks in history, causing widespread damage worldwide. WannaCry led to significant losses in several manufacturing companies, including Nissan and Renault, whose systems were brought down by the virus and were forced to temporarily stop production at several of their car plants across the world.

Kemuri Water Plant—In 2016, malicious actors successfully penetrated a water utility's logic control system and altered the levels of chemicals being used to treat tap water to make it safe for consumption. The group was able to compromise Kemuri Water Company's computers after exploiting web vulnerabilities in its Internet-facing customer payment portal. Sternstein states that the hack "which involved SQL injection and phishing was made easier because log-in credentials for the operational control system were stored on the Web server" [29]. As a result of the hackers gaining access to several databases, personal information of the company's 2.5 million customers also became exposed, and passwords were deemed as compromised as they were unencrypted.

4.4 Fourth Dimension: Production Goals (Table 6)

Low-Impact Vector: Daimler Chrysler Cars—In 2005, a computer worm called Zotob was able to exploit Daimler Chrysler's security vulnerabilities. The Zotob worm affected 13 of the manufacturer's plants. This halted production for periods of time that ranged from 5 to 50 min. The limited production time forced more than 50,000 assembly line workers to stop working between the outages [15]. In

Table 6 Cyber-physical attack events and their impact on production goals

Incident	Attack mechanism	Level of impact	CIA violation	Type of threat	Actors
Daimler Chrysler Cars	Worm	Low	Integrity Availability	Disruption Usurpation	Morocco
Stuxnet	Worm	High	Integrity Availability	Disruption Usurpation	USA and Israel
German Steel Mill	Spear phishing	High	Confidentiality Integrity Availability	Usurpation Disclosure	Unknown
Honda Car Manufacturer	Ransomware	Moderate	Integrity Availability	Disruption Disclosure	North Korea
Taiwan Semi-conductor Manufactur-ing Company	Virus	Moderate	Availability	Disruption	North Korea
Norsk Hydro	Ransomware	High	Confidentiality Integrity Availability	Disruption Disclosure Usurpation Deception	Unknown

2015, Chrysler issued a recall for 1.4 million vehicles that may be affected by a software vulnerability in Chrysler's Uconnect dashboard computers that are prone to be hacked.

Moderate-Impact Vector: Honda Car Manufacturer—In 2017 and 2020, Honda has experienced multiple cyberattacks caused by ransomware. In 2017, the Honda car manufacturing plant that was targeted was located in Japan. The WannaCry virus caused Honda to shut down production in a plant [30]. Jacob Fisher reported that due to an organization-wide network outage, production was halted [31]. It ended up affecting the production of about 1000 cars. Not only did this incident affect the company network in Japan, but also the worm had affected networks across North America, Europe, China, and other regions [32]. In 2020, Honda further experienced another ransomware attack on its network. This attack resulted in shutting down production once again. However, the ransomware also affected the company's "ability to access its computer servers, use email, and make use of its internal systems" [33]. Through the use of ransomware, the hackers were able to encrypt the Honda's data and deny the company's access to their own computers, servers, and IT systems.

Taiwan Semiconductor Manufacturing Company—In 2018, one of the world's largest and most valuable semiconductor companies was infected by a computer virus [34]. This company happens to be a powerful chip manufacturing company and is a supplier of the Apple company. According to Warwick Ashford, "The company said some of its computer systems and 80% of its manufacturing tools had been infected by a virus" [35]. The computer virus caused the production system to be shut down which caused the company to fall behind on its manufacturing of products. As a result, it delayed the shipment of new Apple iPhones [30]. The delay in products not only caused a delay of shipments, but also caused the satisfaction and expectations of customer demands to not be met. One of the most significant impacts of the computer virus attack was the amount of money it cost the Apple company which was estimated at $170 million dollars [36].

High-Impact Vector: Stuxnet in Iran—In 2010, one of the most famous cyber-physical attacks that have ever occurred took place. This attack was caused by a highly developed computer worm which can also be referred to as malware. The worm targeted the Iranian nuclear plant by destroying centrifuges. Hackers were trying to specifically target "the programmable logic controllers (PLCs), collecting information on industrial systems and causing the fast-spinning centrifuges to tear themselves apart" [37]. As a result, a number of centrifuges were destroyed, and it is reported that the attack "may have damaged as much as one-fifth of the nuclear centrifuges" [19]. Stuxnet was a well-executed plan of attack due to the hackers being able to exploit vulnerabilities within the system. By gaining this type of access, the attackers were able to change the displays of the machines to cause misleading readings for the operators [38]. Stuxnet will be remembered for being one of the first successful cyber-physical attacks, as well as causing tremendous physical damage.

German Steel Mill—In 2014, a highly dangerous spear phishing attack was directed towards a steel mill in Germany. The attackers were able to use certain social engineering techniques [19]. By using these attack methods, hackers were able to gain complete access to the plant's networks including the business and production network. This led to failures in multiple system components as well as massive physical damage to the mill [38]. The attackers were able to cause such damage by targeting a blast furnace and disabling its ability to be shut down by administrators and employees [28]. The hackers who planned this attack were well educated on security procedures, advanced systems, and manufacturing process of the steel mill. This attack is important because it caused significant physical damages to the production systems. It is also important to note that if the steel mill had not recognized or stopped the attack, it is possible there could have been serious risks or injuries to employees.

Norsk Hydro—In 2019, one of the world's largest aluminum manufacturing companies was attacked through the use of ransomware. The ransomware that was used for the attack is referred to as LockerGoga. This type of ransomware is designed to "access and encrypt sensitive user data on infected devices, either by sending malicious emails or by using other forms of social engineering" [39]. By infecting 40 sites of the company worldwide [40], hackers were able to gain enough access to shut down both the manufacturer's production system and network. Due to the company's lack of security procedures, all automated manufacturing systems were stopped and were transitioned to being manually operated. This cyberattack had a significant impact on the manufacturing process and financial concerns. The monetary loss from the ransomware attack was "estimated to be between 400 and 450 million NOK, 41–46 million euros, due to loss of production and associated cost" [40]. This equates to between 50 and 54 million US dollars. This can be classified as a high-impact cyberattack because it not only caused the company millions of dollars, but also significantly reduced the production process.

4.5 *Fifth Dimension: Trade Secrets (Table 7)*

Low-Impact Vector: LC Industries—In 2015, a data breach targeted LCI which is a manufacturing organization and technology service provider. The hackers were able to cause a breach that resulted in almost 4000 customer records to vanish [41]. The goal of the attackers was to collect personal information from LCI's customers and consumers. Through the company's investigations, LCI learned that cyberattackers were able to cause the data breach and attain such sensitive information by using malicious code on one of LCI's websites [42]. The information the hackers were able to obtain included customer's names, credit card details, and website log-in credentials. To alleviate the damage from the data breach, LCI shut down the website, removed the malicious code, and implemented new security measures for the website [42]. The impact of this cyberattack is considered low since only a few thousand customers were affected. LCI was also able to detect the attack and make adjustments to provide the website with more secure features.

Table 7 Cyber-physical attack events and their impact on production goals

Incident	Attack mechanism	Level of impact	CIA violation	Type of threat	Actors
Target	Phishing, Malware	High	Confidentiality	Disclosure	Latvia
Havex/Dragonfly	Malware	Moderate	Confidentiality Availability	Disclosure Disruption	Foreign group "Energetic Bear"
LC industries	Malicious code	Low	Confidentiality	Disruption	Unknown
ThyssenKrupp AG	Malware	High	Confidentiality	Disclosure	Southeast Asia
Deloitte Data Breach	Password Breach	Moderate	Confidentiality	Disclosure	Unknown

Moderate-Impact Vector: Havex/Dragonfly—In 2014, a cyber incident called the Havex/Dragonfly occurred and was used to target industrial control systems. It was a type of malware that took the form of a Remote Access Trojan, which compromised such systems as SCADA, PLC, and DCS used within the energy sector [43]. The Dragonfly malware was able to infect hundreds of business computers with the main goal of industrial espionage. The attackers were trying to collect information from as many businesses in the energy sector as possible. The cyberattacks of Havex were executed in an organized and highly skilled approach; it was obvious that the hackers were knowledgeable because the attack was difficult to detect and occurred over a carefully planned time span [44]. Through this, hackers were able to obtain valuable information such as network details, emails, contact lists, and other data. The Havex/Dragonfly was a significant attack because "it was the first advanced attack since Stuxnet to target ICS components" [44].

Deloitte Data Breach—In 2017, Deloitte, which is one of the world's Big Four accounting firms, reported a massive cybersecurity data breach. Hackers were able to gain access to the email server, and this allowed them to obtain valuable information such as usernames, passwords, IP addresses, business information, and workers' health records [45]. This cyberattack led to major security concerns about Deloitte's email system and servers. Evidence showed that the email systems did not have multifactor authentication [46]. This allowed hackers to have easy access to the system because they would just need to figure out one password. Thus, this cyberattack could have been prevented if there were proper multifactor authentication protocols. The impact of the data breach caused employees and clients to lose trust in Deloitte.

High-Impact Vector: Target—In 2013, Target announced that they had discovered an extensive data breach. Hackers were able to gain valuable information about millions of customers. Specifically, the attack "led to the theft of approximately 40 million payment card records along with 70 million other information records such as addresses and phone numbers" [47]. Having customers' information exploited and exposed to people worldwide would cost Target millions of dollars. The company estimated "USD 292 million in incurred expenses as a direct result of the privacy breach, including settlements with four major payment card networks, affected customers, and financial institutions" [28]. Not only did Target lose a large amount of money, but the data breach also caused a decrease in customer sales and performance in their stocks and shares [28]. One can assume that sales decreased because customers lost trust in giving Target their personal information such as credit card information. The impacts caused by this attack had a level of severity which led to Target losing millions of dollars, sales, and ultimately trust of the consumers.

German Company ThyssenKrupp AG—In 2016, one of the world's largest steel makers was targeted by a cyberattack resulting in trade secrets being stolen from the steel production and manufacturing plant design divisions [48]. Hackers gained access to the company's system and obtained a variety of valuable information, such as personal data and design plans. The attackers who organized this malicious plan were knowledgeable and organized. It is believed that the attackers were located in Southeast Asia [49]. The company reported that the attack affected sites in Europe, India, Argentina, and the United States [50]. There were no specific details about how much data, intellectual property, or monetary assets the company lost from the cyber espionage incident. However, we suspect the company could have potentially lost billions depending on what information was stolen.

5 Conclusion

The Internet of Things (IoT) has transformed many aspects of modern smart manufacturing systems. It is evident that smart manufacturing systems are the future of Industry 4.0. However, these systems are much more vulnerable to cyberattacks than traditional manufacturing systems. Given the importance of IoT-based manufacturing systems throughout several industries and economies, identifying and remediating these vulnerabilities are of utmost importance.

To understand potential vulnerabilities and protect manufacturing system standards, this chapter presents a taxonomy of cyberattacks reported in the industry. The proposed taxonomy catalogs and classifies a variety of cyberattacks that have occurred in the past. By using this framework, we hope that both researchers and manufacturers can gain a better understanding of cyberattacks based upon a combination of the NIST's objectives and the potential impact of the attack. The taxonomy will provide the manufacturing industry with a deeper understanding and awareness of the attacks that have occurred on smart manufacturing systems. This will help industries counteract these types of attacks in the future.

As technology evolves and becomes more complex, it will be more challenging to protect industrial control systems (ICS). The number of cyberattacks will increase, and attackers will become more advanced. Cyber security will play an intricate role in developing new techniques to help manufacturers identify, detect, protect, respond, and recover from attacks. In order to protect manufacturing systems, industries will need to pay even more attention to their ICS. Manufacturers will also need to ensure that they are taking the proper security measures and are investing in appropriate secure systems. Companies need to prepare for the worst type of attacks and have incident response, business continuity, and disaster recovery plans in place if and when a cyberattack jeopardizes the organization.

Last year, the Department of Energy funded the Cybersecurity Manufacturing Innovation Institute (CyManII), a $111 million national initiative under the leadership of the University of Texas at San Antonio (UTSA), which consists of national laboratories, universities, manufacturing innovation institutes, industry leaders, and nonprofit organizations. Over the next 5 years, CyManII will help to protect American manufacturing jobs and workers from cyber criminals by becoming more efficient, adaptable, and globally competitive against cyberattacks [51]. The huge investment in this project will assist manufacturers to keep producing products at a high level while also ensuring quality protection of the manufacturing process and systems.

Future research opportunities in this area could include incorporating mitigation techniques for these types of attacks with the NIST standards and framework. Most of these attacks mentioned in the taxonomy could have been prevented if better security protocols were in place. Some of the mitigation methods could include multifactor authentication, encryption, access control, etc. By using this taxonomy, industries can equip themselves with the necessary tools and procedures to protect themselves against such attacks.

Acknowledgements This research was conducted at Tennessee Tech University (TNTech) sponsored by the Office of Research at TNTech and National Science Foundation Scholarship for Service program. We would like to thank Dr. Kim Dukbong for his advice and guidance with securing smart manufacturing systems.

References

1. A. Kusiak, Smart manufacturing. Int. J. Prod. Res. **56**(1–2), 508–517 (2017). https://doi.org/10.1080/00207543.2017.1351644
2. M. Taylor, Sizing up the cybersecurity threat and building the right strategy. *Moldmaking Technology Magazine* 23.3 (2020): 34. Web
3. A. Sebenius, W. Turton, Hacker Boosts Toxic Chemical Level 100-Fold at Water Plant (2021), https://www.bloomberg.com/news/articles/2021-02-08/hacker-changed-chemical-level-in-florida-water-treatment-plant. Accessed 16 Feb 2021
4. A. Ren, et al, *Cyber Security in Smart Manufacturing: Survey and Challenges*, in IIE Annual Conference. Proceedings (2017), pp. 716–721. https://search.proquest.com/docview/1951124648
5. K. Stouffer, et al., Cybersecurity framework manufacturing profile, (NIST Internal Report (NISTIR) 8183) (2019). https://doi.org/10.6028/NIST.IR.8183r1
6. B. Kesler, The Vulnerability of Nuclear Facilities to Cyber Attack; Strategic Insights: Spring 2010, Monterey, California. Naval Postgraduate School (2011), https://explore.openaire.eu/search/publication?articleId=od______2778::4aef096c3896e1e17b98da7330ceba4b

7. A. Greenberg, Hackers remotely kill a jeep on the highway—with me in it (2015), https://www.wired.com/2015/07/hackers-remotely-kill-jeep-highway/. Accessed 20 Oct 2020
8. M. Clayton, Scientists hack into cars' computers—control brakes, engine (2010), https://www.csmonitor.com/USA/2010/0813/Scientists-hack-into-cars-computers-control-brakes-engine. Accessed 20 Oct 2020
9. D. Walker, *TRITON malware threatens industrial machinery with physical damage*. Dennis Publishing Ltd, London (2017), https://search.proquest.com/docview/1984337684
10. M. Thompson, Iranian Cyber Attack on New York Dam Shows Future of War. https://time.com/4270728/iran-cyber-attack-dam-fbi/
11. T. Abdollah, E. Tucker, *Justice Dept. charges 7 in Rye Brook dam cyberattack.*https://abc7ny.com/1260318/
12. T. Abdollah, Banks, dam targeted by Iranian hackers, US says; 7 charged. https://apnews.com/article/20e754ada2eb49edbfe6f5ea823bf5c8
13. E. Montalbano, Florida water plant hack: leaked credentials found in breach database (2021), https://threatpost.com/florida-water-plant-hack-credentials-breach/163919/. Accessed 16 Feb 2021
14. F. Khorrami, P. Krishnamurthy, R. Karri, Cybersecurity for control systems: a process-aware perspective. Dtm, 75–83 (2016). https://doi.org/10.1109/MDAT.2016.2594178
15. R.M. Lee, M.J. Assante, T. Conway, ICS Defense Use Case (DUC) (2014)
16. S. Kim et al., Cyber attack taxonomy for digital environment in nuclear power plants. Nucl. Eng. Technol. **52**(5), 995–1001 (2020). https://doi.org/10.1016/j.net.2019.11.001
17. K. Stouffer, et al, Cybersecurity Framework Manufacturing Profile, (NIST Internal Report (NISTIR) 8183) (2019). https://doi.org/10.6028/NIST.IR.8183
18. E. Byer, M. Fabro, RISI Database (2015). https://www.risidata.com/
19. K.E. Hemsley, D.R.E. Fisher, History of industrial control system cyber incidents. United States (2018). https://doi.org/10.2172/1505628
20. N. Sayfayn, S. Madnick, Cybersafety Analysis of the Maroochy Shire Sewage Spill
21. N. Buckley, H. Kuchler, (−01-05T21:30:50.000Z). Hackers shut down Ukraine power grid. https://www.ft.com/content/0cfffe1e-b3cd-11e5-8358-9a82b43f6b2f
22. A. Shehod, Ukraine Power Grid Cyberattack and US Susceptibility: Cybersecurity Implications of Smart Grid Advancements in the US
23. L.J. Wells, J.A. Camelio, C.B. Williams, J. White, Cyber-physical security challenges in manufacturing systems. Manuf. Lett. **2**(2), 74–77 (2014). https://doi.org/10.1016/j.mfglet.2014.01.005
24. L.D. Sturm, C.B. Williams, J.A. Camelio, J. White, R. Parker, Cyber-physical vulnerabilities in additive manufacturing systems: a case study attack on the .STL file with human subjects. J. Manuf. Syst. **44**(Part 1), 154–164 (2017). https://doi.org/10.1016/j.jmsy.2017.05.007
25. N. Woolf, DDoS attack that disrupted internet was largest of its kind in history, experts say, *The Guardian* (2016). https://www.theguardian.com/technology/2016/oct/26/ddos-attack-dyn-mirai-botnet
26. N. Goud, Massive Cyber Attacks triggers recall of IoT devices in the US (2016). https://www.cybersecurity-insiders.com/massive-cyber-attacks-triggers-recall-of-iot-devices-in-the-us/
27. M. Burgess, Chinese IoT firm recalls 4.3 million connected cameras after giant botnet attack, Wired UK (2016). https://www.wired.co.uk/article/internet-down-dyn-october-2016
28. Anonymous, Types of cyber incidents and losses, in *Enhancing the Role of Insurance in Cyber Risk Management* Anonymous (2017), https://doi.org/10.1787/9789264282148-4-en
29. A. Sternstein, Hackers Manipulate Water Treatment Operations, Invade Adele's Privacy, and Plunder Diamonds on the High Seas (2016), https://www.nextgov.com/cybersecurity/2016/03/hackers-manipulate-water-treatment-operations-invade-adeles-privacy-and-plunder-diamonds-high-seas/126999/. Accessed 20 Oct 2020
30. A. Zarreh et al., Cybersecurity concerns for total productive maintenance in smart manufacturing systems. Procedia Manuf. **38**, 532–539 (2019). https://doi.org/10.1016/j.promfg.2020.01.067

31. J. Fisher, Honda operations disrupted after suspected cyber attack (2020). https://www.bizjournals.com/dayton/news/2020/06/08/honda-operations-disrupted-after-suspected-cyber.html
32. R. Staff, Honda halts Japan car plant after WannaCry virus hits computer network, *Reuters* (2017). https://www.reuters.com/article/us-honda-cyberattack-idUSKBN19C0EI
33. J. Tidy, Honda's global operations hit by cyber-attack, *BBC News* (2020). https://www.bbc.com/news/technology-52982427
34. S. Nellis, D. Shepardson, Taiwan's TSMC to build Arizona chip plant as U.S.-China tech rivalry escalates, *Reuters* (2020). https://www.reuters.com/article/us-usa-semiconductors-tsmc-idUSKBN22Q38T
35. W. Ashford, Virus outbreak at iPhone chip plant could delay shipments. https://www.computerweekly.com/news/252446344/Virus-outbreak-at-iPhone-chip-plant-could-delay-shipments
36. E. White, Apple chip supplier TSMC warns of $170m hit from virus. https://www.ft.com/content/2fe5e096-9909-11e8-9702-5946bae86e6d. Accessed 20 Oct 2020
37. M. Wu, Y.B. Moon, Taxonomy of cross-domain attacks on cyber manufacturing system. Procedia Comput. Sci. **114**, 367–374 (2017). https://doi.org/10.1016/j.procs.2017.09.050
38. A.E. Elhabashy, L.J. Wells, J.A. Camelio, ScienceDirect. J. Environ. Sci. (China) **10**, 125–126 (2017) http://lib.cqvip.com/qk/85265X/201710/7274756650484955494848 4955.html
39. N.E. Oueslati et al, Comparative study of the common cyber-physical attacks in industry 4.0 (2019), https://doi.org/10.1109/IINTEC48298.2019.9112097
40. S. Leppänen, S. Ahmed, R. Granqvist, *Case Study Cyber Security Incident Report—Norsk Hydro* (Springer International Publishing AG, Cham, 2019)
41. J. de Groot, Biggest Manufacturing Data Breaches of the 21st Century. https://digitalguardian.com/blog/biggest-manufacturing-data-breaches-of-the-21-century
42. R. Stallings, *Data Incident Notification* (LC Industries, 2015)
43. N. Tuptuk, S. Hailes, Security of smart manufacturing systems. J. Manuf. Syst. **47**, 93–106 (2018). https://doi.org/10.1016/j.jmsy.2018.04.007
44. N. Nelson, The impact of dragonfly malware on industrial control systems (2016)
45. M. Locklear, Major accounting firm Deloitte reports extensive cybersecurity breach. https://www.engadget.com/2017-09-25-deloitte-reports-extensive-cybersecurity-breach.html
46. J. Goldman, Deloitte suffers major data breach without multi-factor authentication. https://www.esecurityplanet.com/network-security/deloitte-suffers-major-data-breach-without-multi-factor-authentication.html
47. M. Phillips, Target's traffic still hasn't recovered from the giant data breach, Quartz (2014), http://qz.com/212003/targets-traffic-still-hasnt-recovered-from-thegiant-data-breach/. Accessed 25 Oct 2020
48. Reuters, ThyssenKrupp secrets stolen in 'massive' cyber attack (2016-12-08T14:49:16+0000). https://www.cnbc.com/2016/12/08/thyssenkrupp-secrets-stolen-in-massive-cyber-attack.html
49. E. Auchard, T. Käckenhoff, ThyssenKrupp secrets stolen in 'massive' cyber attack, Reuters (2016). https://www.reuters.com/article/us-thyssenkrupp-cyber-idUSKBN13X0VW
50. T. Pultarova, Trade secrets stolen from steel maker ThyssenKrupp in cyber attack. https://eandt.theiet.org/content/articles/2016/12/trade-secrets-stolen-from-steel-maker-thyssenkrupp-in-cyber-attack/
51. U. S. Communications, UTSA officially launches Cybersecurity Manufacturing Innovation Institute (2020), https://www.utsa.edu/today/2020/11/story/cymanii-officially-launches-at-utsa.html. Accessed 16 Feb 2021

Analyzing Performance of Wireless Network Based on the Industrial HART Protocol

M. Kostadinovic, D. Dobrilovic, G. Jotanovic, G. Jausevac, Z. Stojanov, and V. Brtka

1 Introduction

Modern control systems in the process industry increasingly use wireless data transmission to send out information from a sensor located directly at the site of a process to the central control unit. There are numerous reasons for such an approach. The most common reason for choosing a wireless industrial communication network is the cost of installation, since eliminating or significantly reducing cable usage means significant savings in cable and installation work. In other words, the cost of new installation, maintenance, and repair is lower than the corresponding cost for cable solutions, and wireless technology opens up a new field of application in terms of how measured process data is transmitted. Costs in wired networks can be even higher if there is a need to expand existing infrastructure to include additional measurements. While adding new measurement a location in wireless networks represents a small cost and significantly reduces the economic barrier that exists in wired networks. The application of smart manufacturing systems and the expansion of the Industry 4.0 model have created the need to modernize existing industrial environments [1]. One way to modernize existing factories is to use wireless technology in monitoring, controlling, and managing industrial processes.

M. Kostadinovic · G. Jotanovic · G. Jausevac (✉)
Faculty of Transport and Traffic Engineering, University of East Sarajevo, Doboj, Bosnia and Herzegovina
e-mail: miroslav.kostadinovic@sf.ues.rs.ba; goran.jausevac@sf.ues.rs.ba

D. Dobrilovic · Z. Stojanov · V. Brtka
Technical Faculty "Mihajlo Pupin", University of Novi Sad, Zrenjanin, Serbia
e-mail: dalibor.dobrilovic@uns.ac.rs; zeljko.stojanov@tfzr.rs; vbrtka@tfzr.uns.ac.rs

L. Knapčíková, D. Peraković (eds.), *6th EAI International Conference on Management of Manufacturing Systems*, EAI/Springer Innovations in Communication and Computing, https://doi.org/10.1007/978-3-030-96314-9_7

The applicability of WirelessHART technology in the industrial environments, as a wireless technology capable to support system requirements, is investigated in this chapter.

The rest of the chapter is organized as follows: Section 2 gives some insight into related work. Some details on IEEE 802.15.4 standard and WirelessHART are given in Sect. 3. The approach of the WirelessHART system is described in Sect. 4. Section 5 describes the experiment of WirelessHART performance analyses, as well as result discussion, while final Sect. 6 gives the conclusion and possible future work.

2 Related Work

Contemporary industrial systems strive to improve their processes to comply with different technical, safety, and environmental regulations, and to meet corporate financial objectives and strategies. Due to the proven advantages of Industrial Wireless Sensor Networks (IWSNs) compared to traditional wired industrial monitoring and control systems, IWSNs have become essential in modern industrial environments with high flexibility, increased reliability, and self-healing capabilities [2]. Wireless networks with support for flexible use of IoT and sensors in industrial systems are essential in the Industry 4.0 framework [3, 4], leading to intelligent manufacturing systems [5]. Development and implementation of IWSNs in real industrial environments are driven and governed by industrial alliances (manufacturers, vendors, service providers, telecom operators) and governments or regulatory authorities [6]. Güngör and Hancke [2] identified the following challenges for the implementation of IWSNs: (1) resource constraints (energy, memory, and processing); (2) dynamic topologies and harsh environmental conditions; (3) quality-of-service (QoS) requirements; (4) data redundancy; (5) packet errors and variable-link capacity; (6) security; (7) large-scale deployment and ad hoc architecture; and (8) integration with Internet and other networks.

IWSNs with the integration of IoT is a suitable choice for optimizing operational efficiency, automation, and maintenance in industrial processes, where IoT enables large-scale connectivity between machines, computers, and people [7]. Another important use of IWSNs is for monitoring working conditions in industrial environments, aimed at analyzing the impact on worker health and safety [1]. In addition, low-power IWSNs are quite sensitive to electromagnetic noise, which should be considered in designing IWSNs for real industrial applications [8]. Based on the measurement results of electromagnetic noise in an automobile factory, Li et al. [8] calculated that the optimal frequency band for the industrial IoT is 916 MHz. For the measurement and analysis were used log-period antennas and spectrum analyzers.

Design of IWSNs includes work related to hardware development (low-power and low-cost sensor-node development, radio technologies), software development (API, operating system and middleware design, system installation and commissioning), and system architecture and protocol design (network architecture, data aggregation, and fusion, cross-layer design) [2]. Since there are several performance characteristics of complex IWSNs, such as delay or lifetime, depending on their design, it is necessary to optimize and evaluate the network functionality and effectiveness by using experimental methods and simulations [4]. Implementation of IWSNs is highly dependent on standardization efforts, which resulted in several proposed standards in use, such as ZigBee, WirelessHART (an extension of process monitoring and control protocol HART), Ultrawideband (UWB), IETF 6LoWPAN, ISA100, and Bluetooth and Bluetooth Low Energy [2, 4, 9]. Kwon et al. [10] analyzed security issues of the provisioning process in IWSN by researching prominent IWSN standards, such as ISA 100.11a, WirelessHART, and Zigbee. Based on the analysis of security issues through experiments, the authors proposed the measures that can be usable for responding to the security issues.

Hou and Bergmann [9] presented an industrial wireless sensor network (IWSN) for monitoring industrial machine conditions and diagnosing faults. Experimental IWSN supports on-sensor feature extraction and fault diagnosis and Dempster–Shafer classifier fusion. The system is based on three sensor nodes, one coordinator node, one laptop, and LabVIEW software. The proposed system was tested by a set of laboratory experiments on a single-phase induction motor. The motor vibration was measured by an ADXL335 microelectromechanical system (MEMS) accelerometer, while the stator current signal is monitored by a current transformer DIGITECH OM-1565. The JN5139 microcontroller with the support for IEEE 802.15.4 and ZigBee protocols was selected as a sensor node suitable for on-sensor data processing. Experimental results revealed that on-sensor fault diagnosis could reduce the payload transmission, decrease the node energy consumption, and prolong the lifetime of nodes.

An implementation of WirelessHART in the NS-2 network simulator is presented by Zand et al. [11]. The implementation supports both the WirelessHART network manager and the whole stack of OSI layers of the WirelessHART standard. The simulator also includes a security layer to enable secure and reliable communications. For the simulation, a sniffed traffic from a real WirelessHART testbed installed in the Idrolab plant in Italy was used. A testbed is designed in a way to present an instrumented steam generation process within a real industry environment. Empirical results indicate that the results obtained during the simulation are quite close to the results from the real networks.

A general customized supervision model for industrial assembling systems that incorporates wireless sensor network (WSN) technology for assembling process management is presented in [12]. The main contribution of the model is the solution to the problem of localization during the assembling process. Artificial neural networks and weighted k-nearest neighbor method are used in localization algorithms. The proposed supervisor model for WSN contains four layers: (1) production layer with the wireless sensors in the assembling system, (2) preparatory layer is used for data transfer by using wired and wireless communication via standard communication protocols, (3) control layer evaluates measured values and stores them in a database, and (4) management layer produces reports based on stored values in the database. Sensors in WSN use ZigBee technology based on

IEEE 802.15.4 standard for communication. The proposed model is tested in an industrial laboratory that simulates the real industrial environment.

3 WirelessHART and 802.15.4 Overview

WirelessHART is a wireless sensor networking technology based on the Highway Addressable Remote Transducer Protocol (HART). It is designed as a multi-vendor, interoperable wireless standard in accordance with the requirements of process field device networks. WirelessHART is ratified in September 2007 by the HART Communication Foundation, as a part of the HART Field Communication Specification, Revision 7.0. It is the first standard specifically targeting industrial applications, process measurement, and control applications. WirelessHART uses a multi-hop full mesh network topology and time-division multiple access (TDMA) as the channel access method. With TDMA, the network communication is divided into guaranteed time slots (GTS) 10 ms long, where each GTS is reserved for a specific communication link ensuring contention-free utilization of the radio channel. The duration of the time slot provides enough time for the transmission of one data packet from the source device and the acknowledgment from the recipient of the data packet. WirelessHART is based on the PHY and MAC layers of the IEEE Std 802.15.4 [13].

IEEE 802.15.4 is a technical standard that defines the operation of low-rate wireless personal area networks (LR-WPANs). It specifies the physical layer and media access control for LR-WPANs and is maintained by the IEEE 802.15 working group, which defines the standard in 2003. This standard is used as the basis for a group of network standards and protocols designed for wireless sensor networks such as Zigbee, ISA100.11a, WirelessHART, 6LoWPAN [14], and 802.15.4 g Wi-SUN [15]. Those standards offer a broad range of applications such as wireless sensor networks (WSN), industrial WSN, smart home, smart metering, and smart utility.

The IEEE Std 802.15.4 defines the PHY and the MAC sublayer for low-rate wireless personal area networks. The standard specifies PHY as it is shown in Fig. 1. The standard defines 27 channels, numbered 0–26. Channel 0 is in the 868 MHz band with a center frequency of 868.3 MHz, channels 1–10 are in the 915 MHz band with a channel spacing of 2 MHz, and channel 1 is with a center frequency of 906 MHz. Channels 11–26 are in the 2.4 GHz band. The channel spacing is 5 MHz and the center frequency of channel 11 is 2.405 GHz [16]. WirelessHART operates in 2.4 GHz and does not use channel 26.

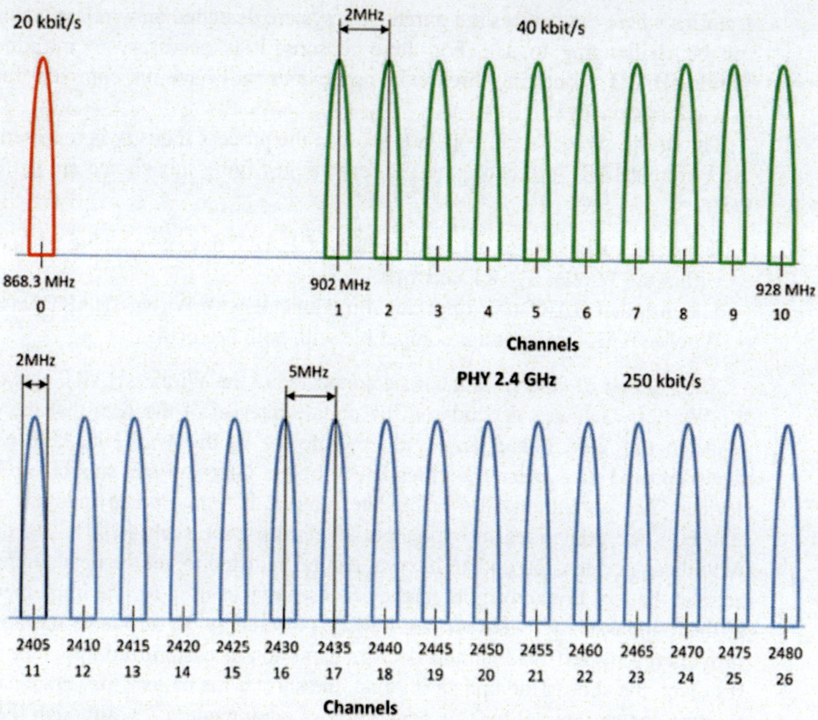

Fig. 1 IEEE 802.15.4 spectrum usage

4 WirelessHART System Architecture

The world's leading manufacturers of process measuring equipment have focused on the problems in industrial wireless sensor networks [17]. The result is the application of wireless communication technology based on the HART protocol, which can lead to significant cost reductions in production processes [18]. That is, when this technology is integrated into an industrial process, it must provide almost limitless possibilities ranging from device monitoring and diagnostics to production process management. However, in the field of industry, applications with very strict reliability requirements, such as security and time-critical functions, have not been implemented in WirelessHART systems [19]. Therefore, the WirelessHART systems become challenging for the study.

Considering the previous achievements, in this case, a model of the industrial communication network (ICN) using WirelessHART network was proposed and supported [20]. The WirelessHART model is particularly suitable for the introduction of an industrial wireless communication network (IWCN) in factories under construction and factories with custom infrastructure. In our region, there are mostly factories where the devices are wired, so a system designed on wireless technology can be challenging to use. For these reasons, it is necessary to introduce the WirelessHART model in which existing devices will have the complete function of wireless devices [21, 22].

The future choice of network protocols in the process industry is represented by the WirelessHART protocol, and the reasons justifying this choice are as follows [23]:

- It provides support for a large number of HART devices capable of operating within the WirelessHART network.
- The installed HART networks can communicate with WirelessHART networks.
- WirelessHART devices are compatible with both protocols.

The number of devices that can be connected to the WirelessHART network via the Wireless Gateway depends on the update interval of the data that the device sends to the WG. For example, if each device in the WirelessHART network sends updated data every 60 s then the Wireless Gateway can support up to 100 devices. The maximum allowed update interval is 1 h. Update intervals shorter than 60 s are predefined in exponents of number two, such as 4, 8, 16, or 32 s. With these devices, the update interval can be adjusted to get the best compromise between battery life and signal latency. These devices used for the implementation of the WirelessHART network are battery powered (9 V) and have the ability to save energy by self-reading and data transmission at customized update intervals. However, the shorter the update interval, the shorter the battery life. Practically the shortest update interval for this generation of equipment is 8 s, although there is a choice of 4, 2, or 1 s. These devices have the ability to estimate the remaining days of battery life based on their current state of charge and power consumption.

In the closed space of the industrial environment, there are obstacles that are on the line of communication between the devices within the realized wireless network. The enclosed space architecture with the marked Wireless Gateway position and the numbered device positions wireless transmitter (node) together with the indicated distances between them were organized for the experiment.

The WirelessHART architecture shown in Fig. 2 contains the following elements:

- HOST System
- Wireless Gateway (WG)
- Wireless transmitter (WT)

4.1 Host System

The Host system consists of the Host computer and appropriate software applications that allow monitoring, control, and management of industrial processes. Data collected with the help of WirelessHART does not always belong to the

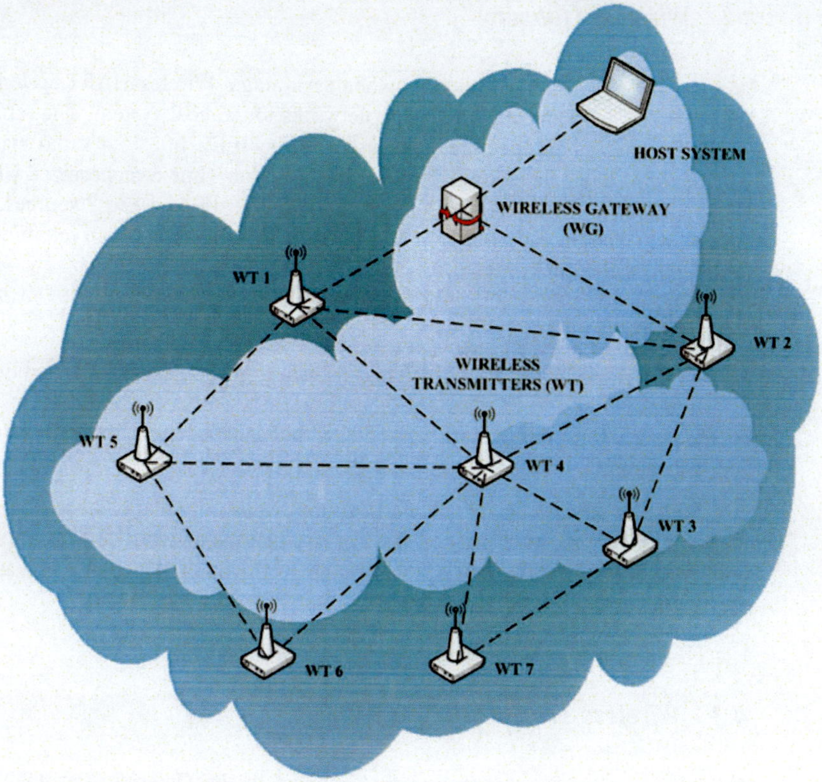

Fig. 2 WirelessHART architecture

control system, as some measurement points can only be used for monitoring or management. They can also be used for real-time control in certain applications. Host systems should be such that data packet from the WirelessHART network is delivered to the Host computer via the Gateway.

WirelessHART data can be routed through shared network resources:

- Data collected on liquid level in industrial tanks
- Data collected on the vibration spectrum
- Data collected on fluid pressure
- Data collected on the temperature state of the observed industrial system, etc.

A well-designed Host system ensures efficient collection of data from WirelessHART devices and proper dissemination of this information to the Host computer. Host computers use databases and specific applications to acquire data. With the help of integrating WirelessHART data via standard interface protocols, databases can be automatically populated with information from WirelessHART devices.

4.2 Wireless Gateway

Wireless Gateway is very important when designing a WirelessHART system. The WG connects wireless self-organizing networks to the host system. The reliability of WG's wireless architecture is ensured by the IEEE 802.15.4 radio standard, 2.4 GHz ISM band sliced into 15 radio channels with Time Synchronized Channel Hopping to avoid interference from other radios, Wi-Fi, and EMC sources. Also, Direct Sequence Spread Spectrum (DSSS) technology delivers high reliability in a challenging radio environment.

Important characteristics of the WG are:

- Easy configuration and management of self-organizing networks
- Easy integration into control systems and data applications via serial and Ethernet connections
- Wireless capabilities that are expanded with the advantage of using in inaccessible locations

Layered security keeps the network secure by ensuring that data transmission is received only by the WG. Also, it is necessary that the network devices implement industry-standard Encryption, Authentication, Verification, Anti-Jamming, and Key Management.

4.3 Wireless Transmitter

Wireless transmitters are devices that connect to the Gateway and enable fast, reliable, and secure data transfer. Often, the WT includes sensors that gather information from industrial environments. WirelessHART devices include wireless transmitters with temperature, pressure, flow, level, and other process variable sensors.

WT in the wireless network offering dual channel, discrete inputs, and discrete outputs:

- Discrete single- or dual-switch input with logic for limit contact and opposing contact applications.
- Dual channels are each configurable for discrete input or discrete output.

The connection between the WT to each other and the connection of the WT to the WG are essential for data transfer in the Host system.

5 Results and Discussion

The experiment took place in a laboratory environment. The experiment setup is as follows: the Wireless Gateway is placed on the wall in the far-left wing of the laboratory as it is shown in Fig. 3. The wireless transmitter (node 2) is placed in the fixed location near the Wireless Gateway. The wireless transmitter (node 1) is portable, and it is used to test the WirelessHART network performance at four different distances of 5, 10, 20, and 30 m.

The results of the tests are presented in the following tables. The overall network statistic is given in Table 1. Each location displays the results for node 1, node 2, and WG, respectively. The results are for the number of neighbors connected, reliability of the links, missed updates, path stability, received signal strength indicator (RSSI), and joins. The results give the similar values of RSSI for different locations, giving the conclusion that according to the results, the distance in this area of coverage does not have significant influence on the RSSI level.

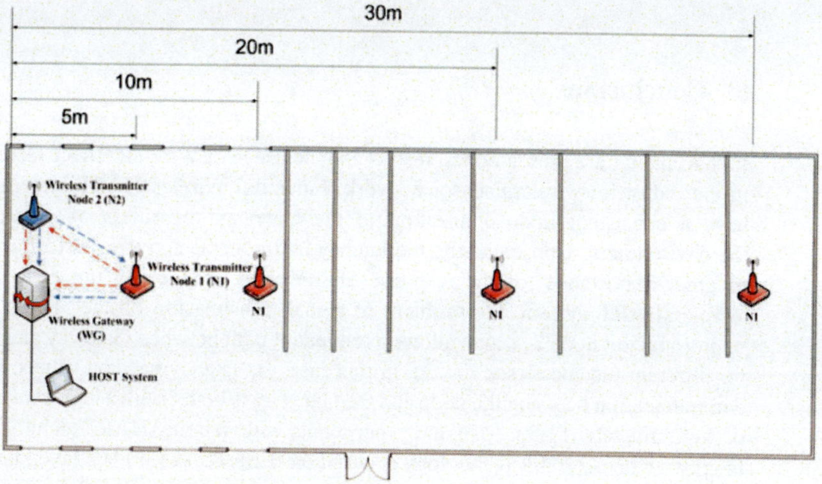

Fig. 3 WirelessHART experimental network layout

Table 1 Network device status

	5 m			10 m			20 m			30 m		
	N1	N2	WG	N1	N2	WG	N1	N2	WG	N1	N2	WG
Neighbors	2	2	/	2	2	/	2	2	/	2	2	/
Reliability (%)	100	99.9	/	100	100	/	100	100	/	95,9	99.9	/
Missed updates	0	2	/	0	0	/	0	0	/	736	158	/
Path stability (%)	100	100	/	100	100	/	100	100	/	100	100	/
RSSI (dB)	−42	−32	/	−41	−30	/	−40	−28	/	−41	−28	/
Joins	1	1	/	1	1	/	6	1	/	6	1	/

Table 2 shows the results for the node 1 statistics. Here, the results are focused on RSSI values in direction to and from the neighbor. The results are slightly better for the distance of 5 m, and slightly worse but almost equal for other distances. The path stability is almost 100% in each scenario.

Table 3 shows the results for the node 2 statistics. Here, the results are almost simi-lar as Table 2 for node 1. The only difference here is the better results for 30 m of node 1 signal strength compared to shorter distances. This small deviation from the expected results can be explained with the institutional (building) layout, wall, furniture, and other obstacles.

Table 4 shows the results for the burst statistic with the focus on latency in the network. The latency here gives very good results. This is of high importance for planning possible applications of the WirelessHART in the given system, because the latency is crucial for optimal network performance in a given system.

Table 5 shows the results indicating the overall network performance. Here the most important parameter is the number of burst messages, showing the intensity of network traffic. Again, average latency in 15-min interval and latency lifetime are very good for potential usage of the system.

6 Conclusions

In this chapter are presented test results and the usage of WirelessHART technology in the industrial wireless sensor network. Industrial WirelessHART systems must have a consistent node connection to the Gateway regardless of obstacles in the environment. Consequently, the latency and general network performance are of great importance for the response of the entire system. In the experimental WirelessHART system, the problem of signal transmission obstacles was solved by introducing node 2, which allows a redundant path between Gateway and node 1 for different distances (see Fig. 3). In this case, the latency has low values, and the communication between the node and the Gateway is better with a low-transmission packet error rate (PER). Also, the experiments with WirelessHART technology are interesting because this technology is based on IEEE 802.15.4 PHY layer, designed especially for WPANs. This proved to be important for ZigBee, WirelessHART, 802.15.4 g SUN, 6LoWPAN, etc.

The experiments took place in a laboratory environment with three nodes. Two nodes were fixed (WG and WT as node 2) and one node was portable (WT as node 1). Results and discussion presented in Sect. 5 justify the applicability of WirelessHART technology in the industrial environments as a wireless technology capable to support system requirements. The WirelessHART with its features such as low latency and low PER/low number of errors on the specified distances shows to be highly applicable in the system for the industrial monitoring. The most important performance parameter analyzed in this research is latency. The low latency values justify the applications of the WirelessHART in the given system because the latency

Table 2 Node 1 statistics

	5 m			10 m			20 m			30 m		
	N1	N2	WG	N1	N2	WG	N1	N2	WG	N1	N2	WG
Network joins	1			1			6			6		
Neighbor count	2			2			2			2		
Rcv signal strength to neighbor (dB)	/	−47	−43	/	−52	−41	/	−53	−40	/	−53	−43
Rcv signal strength from neighbor (dB)	/	−48	−42	/	−53	−41	/	−53	−43	/	−54	−42
Path stability (%)	/	100	99.6	/	100	100	/	100	100	/	100	100

Table 3 Node 2 statistics

	5 m			10 m			20 m			30 m		
	N1	N2	WG	N1	N2	WG	N1	N2	WG	N1	N2	WG
Network joins	1			1			1			1		
Neighbor count	2			2			2			2		
Rcv signal strength to neighbor (dB)	−48	/	−33	−53	−48	/	−33	−53	−48	/	−33	−53
Rcv signal strength from neighbor (dB)	−47	/	−30	−52	−47	/	−30	−52	−47	/	−30	−52
Path stability (%)	100	/	99.3	100	100	/	99.3	100	100	/	99.3	100

Table 4 Burst statistics

	5 m			10 m			20 m			30 m		
	N1	N2	WG	N1	N2	WG	N1	N2	WG	N1	N2	WG
Last burst message latency (s)	0.028	0.060	/	0.761	0.083	/	1100	0.282	/	0.688	0.678	/
Average burst message latency (s)	0.369	0.282	/	0.231	0.227	/	0.760	0.415	/	0.700	1031	/
Minimum burst message latency (s)	0.004	0.027	/	0.000	0.001	/	0.048	0.028	/	0.001	0.540	/
Maximum burst message latency (s)	1883	0.950	/	1766	0.658	/	2489	1350	/	2665	1968	/

Table 5 Network statistics

Parameter	5 m	10 m	20 m	30 m
Tx requests	6	0	11	11
Tx request timeouts	0	0	0	0
Rx response messages	6	0	11	10
Rx burst messages	3289	3447	3555	3864
Requests received	6	0	11	11
Responses sent	6	0	11	10
Upstream packets lost	0	0	0	0
Average latency 15 min	0.586 s	0.700 s	0.497 s	0.519 s
Average latency lifetime	0.675 s	0.671 s	0.524 s	0.524 s

and better network performance in general are crucial for the fast response required in this system.

In future work, the coexistence of the WirelessHART with other wireless technologies operating in the 2.4 GHz band, e.g., IEEE 802.11 and Bluetooth Low Energy, should be examined as well.

References

1. D. Dobrilovic, V. Brtka, Z. Stojanov, G. Jotanovic, D. Perakovic, G. Jausevac, A model for working environment monitoring in smart manufacturing. Appl. Sci. **11**, 2850 (2021)
2. V.C. Gungor, G.P. Hancke, Industrial wireless sensor networks: challenges, design principles, and technical approaches. IEEE Trans. Ind. Electron. **56**, 4258–4265 (2009)
3. X. Li, D. Li, J. Wan, A.V. Vasilakos, C.-F. Lai, S. Wang, A review of industrial wireless networks in the context of industry 4.0. Wirel. Netw **23**, 23–41 (2017)
4. Y.-S. Yu, Y.-S. Chen, A measurement-based frame-level error model for evaluation of industrial wireless sensor networks. Sensors **20**, 3978 (2020)
5. J. Wan, B. Chen, M. Imran, F. Tao, D. Li, C. Liu, S. Ahmad, Toward dynamic resources management for IoT-based manufacturing. IEEE Commun. Mag. **56**, 52–59 (2018)
6. Z. Sheng, C. Mahapatra, C. Zhu, V.C. Leung, Recent advances in industrial wireless sensor networks toward efficient management in IoT. IEEE access. **3**, 622–637 (2015)
7. D.E. Boubiche, A.-S.K. Pathan, J. Lloret, H. Zhou, S. Hong, S.O. Amin, M.A. Feki, Advanced industrial wireless sensor networks and intelligent IoT. IEEE Commun. Mag. **56**, 14–15 (2018)
8. M. Li, W. Gu, W. Chen, Y. He, Y. Wu, Y. Zhang, Smart home: architecture, technologies and systems. Procedia Comput. Sci. **131**, 393–400 (2018)
9. L. Hou, N.W. Bergmann, Novel industrial wireless sensor networks for machine condition monitoring and fault diagnosis. IEEE Trans. Instrum. Meas. **61**, 2787–2798 (2012)
10. S. Kwon, J. Jeong, T. Shon, Toward security enhanced provisioning in industrial IoT systems. Sensors **18**, 4372 (2018)
11. P. Zand, E. Mathews, P. Havinga, S. Stojanovski, E. Sisinni, P. Ferrari, Implementation of wirelesshart in the ns-2 simulator and validation of its correctness. Sensors **14**, 8633–8668 (2014)
12. L. Gogolák, I. Fürstner, Wireless sensor network aided assembly line monitoring according to expectations of industry 4.0. Appl. Sci. **11**, 25 (2021)

13. 802.15 WG - Wireless Specialty Networks (WSN) Working Group: IEEE Standards Interpretation for IEEE Std 802.15.4™—2006 IEEE Standard for Information technology—Telecommunications and information exchange between systems—Local and metropolitan area networks—Specific require-ments—Part 15.4: Wireless Medium Access Control (MAC) and Physical Layer (PHY) Specifications for Low-Rate Wireless Personal Area Networks (WPANs) (2006), https://standards.ieee.org/standard/802_15_4-2006.html#Working
14. G. Montenegro, N. Kushalnagar, J. Hui, D. Culler, Transmission of IPv6 packets over IEEE 802.15. 4 networks. Internet Propos. Stand. RFC. **4944**, 130 (2007)
15. IEEE Standard for Local and metropolitan area networks—Part 15.4: Low-Rate Wireless Personal Area Networks (LR-WPANs) Amendment 3: Physical Layer (PHY) Specifications for Low-Data-Rate, Wireless, Smart Metering Utility Networks. IEEE Std 802.15.4g-2012 (Amendment to IEEE Std 802.15.4-2011). 1–252 (2012), https://doi.org/10.1109/IEEESTD.2012.6190698
16. B.M. Wilamowski, J.D. Irwin, *Industrial Communication Systems* (CRC Press, 2016)
17. P. Soldati, H. Zhang, M. Johansson, Deadline-constrained transmission scheduling and data evacuation in WirelessHART networks. Presented at the 2009 European Control Conference (ECC) (2009)
18. International Electrotechnical Commission, *Industrial Communication Networks-Wireless Communication Network and Communication Profiles-WirelessHART* (Geneva, Switzerland, 2010)
19. J. Song, A.K. Mok, D. Chen, M. Nixon, Challenges of wireless control in process industry. Presented at the Workshop on Research Directions for Security and Networking in Critical Real-Time and Embedded Systems (2006)
20. T.P. Raptis, A. Formica, E. Pagani, A. Passarella, On the performance of data distribution methods for wireless industrial networks. Presented at the 2019 IEEE 20th International Symposium on "A World of Wireless, Mobile and Multimedia Networks" (WoWMoM) (2019)
21. Pratt, W.A., Nixon, M.J., Rotvold, E.D., Pramanik, R.S., Lennvall, T.P.: Combined Wired and Wireless Communications with Field Devices in a Process Control Environment. (2014)
22. J. Åkerberg, M. Gidlund, T. Lennvall, J. Neander, M. Björkman, Efficient integration of secure and safety critical industrial wireless sensor networks. EURASIP J. Wirel. Commun. Netw., 1–13 (2011, 2011)
23. X. Jin, A. Saifullah, C. Lu, P. Zeng, Real-time scheduling for event-triggered and time-triggered flows in industrial wireless sensor-actuator networks. Presented at the IEEE INFOCOM 2019-IEEE Conference on Computer Communications (2019)

Modeling the Purchase Process of the InnIoTShop Solution in a Store Environment

Dragan Peraković, Marko Periša, Ivan Cvitić, and Petra Zorić

1 Introduction and Research Motivation

The quality of life of people with visual impairment is impaired by various obstacles they face daily. The independence of these persons is possible with assistive technology, development, and application, which leads to the easier overcoming of obstacles. Assistive technologies define or improve existing solutions in order to enable the independence of people with specific disabilities. Smart shopping carts for blind and partially sighted people are one of the examples of innovative solutions for assistive technologies.

The activity of self-shopping for people with visual impairments with the help of a white cane and a guide dog is further complicated since most stores do not allow dogs to move around the food departments [1]. In the 1990s, "purchasers" were first mentioned at the time of purchase, representing scanners, and they functioned in such a way as to fit the desired item and read the price and characteristics of the product [2]. In 2016, smart shopping carts for the visually impaired appeared for the first time in India, which had a tablet in their offer that displayed the prices of products and their characteristics. For many visually impaired people, the price shown made it much easier to buy because they knew the amount of the product in advance [3]. Information and communication technology in the function of assistive technologies to increase the mobility of people with visual impairment during everyday shopping is presented in [4].

Scientific research dealing with innovative information and communication services and systems in a smart trade environment is increasingly focusing on sensor

D. Peraković · M. Periša · I. Cvitić · P. Zorić (✉)
Faculty of Transport and Traffic Sciences, University of Zagreb, Zagreb, Croatia
e-mail: dragan.perakovic@fpz.unizg.hr; marko.perisa@fpz.unizg.hr; ivan.cvitic@fpz.unizg.hr; petra.zoric@fpz.unizg.hr

L. Knapčíková, D. Peraković (eds.), *6th EAI International Conference on Management of Manufacturing Systems*, EAI/Springer Innovations in Communication and Computing, https://doi.org/10.1007/978-3-030-96314-9_8

technologies and networks. In [5], the development of smart carts based on RFID and ZigBee technologies and networks was proposed. Reducing queue delays using the electric smart trolley system is the main focus of the work [6], while the Bascart system is also based on the help of RFID tags on products to speed up the billing process in stores [7]. The smart cart in the store accompanies the user as he/she performs the purchase process, and the automatic billing system gives the user the ability to view the total amount of the invoice [8].

However, a much smaller number of scientific research focused on developing innovative solutions for the target group of users of this work, i.e., people with visual impairment. An overview of the various technologies that have been developed to help this group of users in performing the purchase process is the topic of the chapter [9]. Portable devices that can be used when purchasing with a focus on energy consumption, weight, ease of use, and economy are discussed [10]. A cart-mounted system has been proposed, which uses the concept of return of investment (ROI), text localization, and recognition using special software and a webcam [11].

The development of innovative information and communication services enables the provision of all relevant information to the business process stakeholders [12]. Business model innovations are crucial to a company's ability to achieve growth and long-term sustainability [13]. Many business model innovations are not sustainable, leading to severe economic complications for companies developing specific solutions [14]. When establishing a business model, it is imperative to take care of its sustainability [14].

Currently available scientific papers are mainly based on developing information and communication services/products/systems for the visually impaired conceptual or prototype level. The authors of this chapter have noticed that these papers do not have developed business models for specific solutions that they develop.

2 InnIoTShop System Architecture and Value Chain

The conceptual architecture of the InnIoTShop service delivery system is shown through layers in Fig. 1. On the perception layer, sensor nodes route data through the network layer using access point (AP) and gateway (G) to Fog, or cloud computing and storage. The data collected in the sensor nodes are located on the sensors installed on the shopping cart and in the shopping store. In the middleware layer, data collection, fusion, processing, and storage of collected data occur. The billing process occurs in the business layer, while the necessary application solutions for the InnIoTShop service to end users are located in the application layer.

For this chapter, a company called InnIoTSolution that develops the InnIoTShop service is defined. Service providers can be divided into network infrastructure service providers (network provider) and content service providers, in this case, InnIoTSolution. The network operator provides the necessary network infrastructure for service delivery (3G, 4G, 5G, Internet). The required user equipment is a mobile

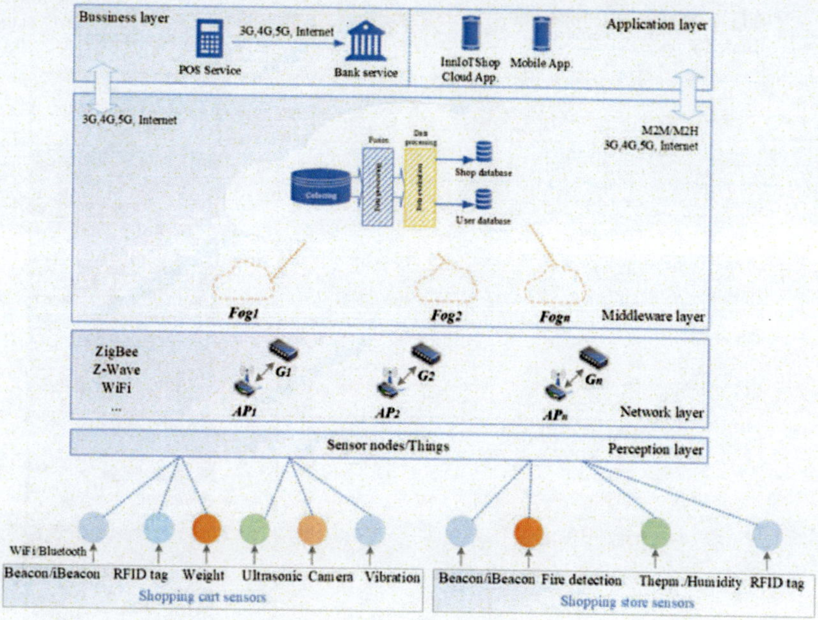

Fig. 1 Conceptual architecture of the InnIoTShop system [15]

device on which a mobile application is implemented that provides access to the necessary information and whose design contains universal design elements.

The functionalities offered by this solution can be divided into functionalities that are available through the mobile application and the functionality of the cloud infrastructure application, which are collectively called the functionalities of application solutions [15]. All functionalities are executed in real time. Some of the suggested functionalities are creating a list and ordering process, content-dependent data storage or product detection, and checking the accuracy of the selected product.

The equipment located in the InnIoTStore within which the InnIoTShop service is performed contains several information and communication subsystems that interact with

$$S_{\mathrm{i}} = \left\{S_{\mathrm{c}} + S_{\mathrm{app}} + S_{\mathrm{cc}}\right\} \tag{1}$$

where S_{i} represents equipment, S_{c} represents smart cart subsystem, S_{app} represents application subsystem, and S_{cc} represents cloud computing subsystem. There is also the possibility of expanding the system by adding new services for the end user.

The concept of service offering is based on the cloud computing concept, whose purpose is to provide everything as a service (XaaS). There are three basic types of service models: Infrastructure as a Service (IaaS), Platform as a Service (PaaS), and Software as a Service (SaaS). Each of these models differs according to the type

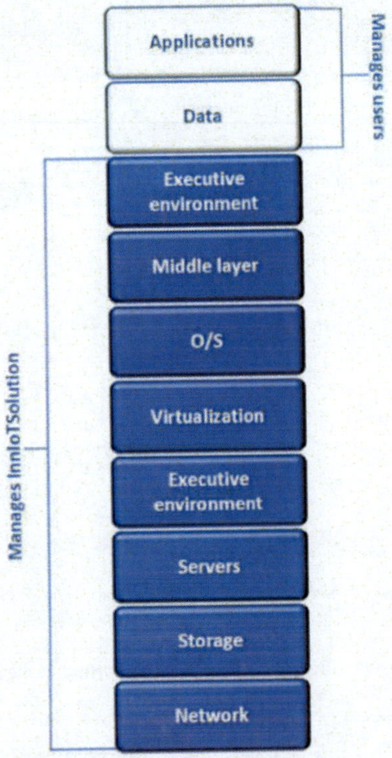

Fig. 2 PaaS service model of InnIoTShop

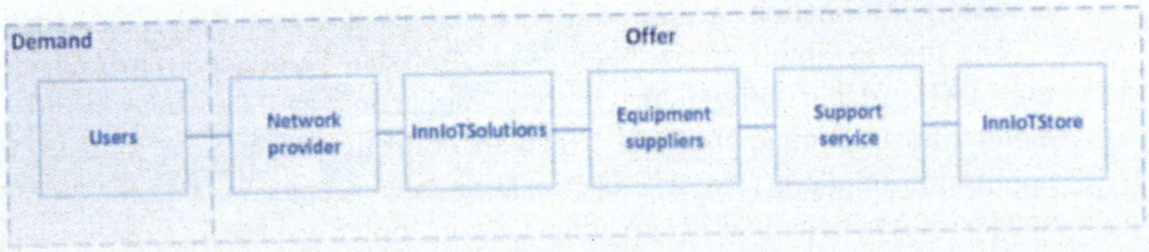

Fig. 3 InnIoTShop service value chain

and range of services it offers. In this example, InnIoTSolution develops the PaaS service model of InnIoTShop. The services available in the PaaS model are shown in Fig. 2.

The basis for developing a business model is a value chain that describes products, services, and relationships, and each chain participant adds new value before handing over the activity to the next chain participant. The value chain explains how business models are created and developed: on one side of the value chain, on the example of the InnIoTShop service, demand, i.e., end users, and on the other side, offer, as shown in Fig. 3.

In the business process proposal from this chapter, the following stakeholders of the value chain of delivery of the InnIoTShop service are defined:

- User
- Network provider
- InnIoTSolution
- Equipment suppliers
- Support service
- InnIoTStore

The user represents a blind or partially sighted person who uses the proposed InnIoTShop service to perform purchasing activities within the custom InnIoTStore. The user can also be a person without disabilities and/or an elderly and slow-moving person. The network provider enables the use of network data traffic technology for InnIoTShop users and InnIoTStore that are introducing the solution due to the impossibility of using the solution without connecting to the Internet network. InnIoTSolution is a company working on the development, implementation, and maintenance of InnIoTShop solutions. Equipment suppliers are companies that enable the procurement of the necessary equipment for the development of solutions. Support service is responsible for resolving difficulties that may arise when using the solution and is a separate part of InnIoTSolution's business. InnIoTStore is an intelligent store that offers carts equipped with a network of sensors and actuators.

Each of these stakeholders can be assigned roles that specify the meaning of each stakeholder within the transformation accelerator tool. For example, the user stakeholder is assigned the role of consumer, while the employee's role is assigned to the stakeholder support services. InnIoTSolution has the role of solution provider, and equipment suppliers have the role of procuring the equipment needed to make smart carts. The support service acts as a consultant and solution provider in the InnIoTStore, which acts as a solution provider to the end user.

3 Business Scenario

It is possible to define three business scenarios:

- User and mobile application scenario (A_{uma})
- Smart store scenario containing smart cart and mobile application (A_{scm})
- Cloud computing infrastructure scenario in which the focus is on mobile application and cloud infrastructure application (A_{cci})

The information and communication system (A_{ic}) for the delivery of the InnIoTShop service can provide the possibility of service delivery in three different scenarios, and is made up of a union of these scenarios

$$A_{ic} = A_{uma} \cup A_{scm} \cup A_{cci} \tag{2}$$

or if there is expansion with additional scenarios

$$A_{ic} = \cup_{(i=1)}^{A} Ai \tag{3}$$

According to the functionalities of the transformation accelerator tool, for the development of the business scenario of the InnIoTShop service, with previously defined stakeholders in the value chain of service delivery, it is necessary to identify added values and key drivers and make a SWOT analysis of InnIoTShop service development. The added value relates to the innovation that InnIoTSolution promises to provide to the users of their service if users choose to use it. The key drivers are the leading factors influencing the success of InnIoTSolution or its business. The SWOT analysis of the development of the InnIoTShop service is essential due to the development of the business strategy InnIoTSolution.

3.1 Value Proposition

Depending on the business scenario being observed, the added value can be divided into three parts. For A_{uma}, the added value is the accuracy of the information, speeding up the purchase process, user independence, improving the quality of life, adapting content, and safe movement. For A_{scm}, the added value is market competition, use of new technology, acceleration of the purchasing process, socially responsible business, and new services to smaller groups of users. For the A_{cci}, these are content customization, real-time information, transaction security, ease of application maintenance, and modularity.

Accuracy of information refers to obtaining correct and timely feedback on the environment in which the user is located and obtaining feedback during the product search process. It is a value of great importance for the user of the solution because it increases the user's sense of security when using the solution, leading to better acceptance of the solution by the user. Speeding up the buying process is a value that saves user's time. Performing specific activities in the purchase process is reduced, and the amount of time that the user will spend in the store. Independence in purchasing represents the independence of the user of the solution from the help of other persons, which leads to an increase in the user's ability and an improvement in the quality of life. The possibility of independent performance of life activities can thus lead to an improvement in users' quality of life, which is a value that leads to better acceptance of the proposed solution.

Value can also be achieved by providing an innovative solution on the market, i.e., by offering a solution adapted to the needs of smaller groups of users with disabilities whose needs are often ignored by those working on the development of new solutions. Corporate social responsibility enables companies to work on introducing and developing such a solution to promote their business on the market further to affect society, i.e., members of the community positively.

The content provided to users is adapted to their needs, thus creating the added value of the InnIoTShop solution itself. Providing real-time information leads to the security of the users themselves and speeds up the purchase process. Every transaction that takes place in InnIoTShop is protected in different ways, which creates security for users. Because the applications are located in the cloud, they are easy to maintain and update, and the modularity of the service is visible when registering users in the system as it can be used by visually impaired, non-visually impaired, and/or elderly and slow-moving people.

3.2 *Key Drivers*

The key drivers are the leading factors influencing the success of InnIoTSolution or its business. They can thus have a big impact on whether a company is doing well or not, and they can also point to problems of poorer performance or business results. The key drivers of solution development are:

- Development of technology
- Customer satisfaction
- ROI
- Company growth
- Hiring new people
- Costs
- Corporate social responsibility
- Modularity of the solution

The development and increased use of innovative technologies, devices, and networks, considering, in this case, mobile devices, mobile networks, and IoT technologies, will enable InnIoTSolution to develop new innovative solutions. They will be able to create added value and improve the InnIoTShop service.

Customer satisfaction with the provided solution is one of the key drivers for further development of the solution. If the users are not satisfied with the proposed solution, there is a possibility of refusing to use it. In this case, the company that develops the solution is at a loss, and it is imperative to include people with visual impairments in designing and creating the solution. In this way, feedback occurs that directly affects the final result. This is possible with associations that deal with the difficulties that people with visual impairments face daily. Their engagement from the beginning of designing the service until the creation of the final product is significant because of the user requirements that can help the project team define certain functionalities.

Return on investment is crucial for investors who invest in it at the very beginning of the service development because of the opportunity to make a financial profit in the future. The company's growth in the market provides the opportunity to hire new people who can contribute to new ideas and skills, which further improves the business and accelerates the development of solutions. The costs incurred in creating and maintaining the service are significant from the financial point of view of InnIoTSolution's stability. Socially responsible business is one of the most critical drivers of solution development. The InnIoTShop solution is intended primarily for visually impaired people who belong to a small group of users in society who face obstacles in performing daily activities. InnIoTShop solution is modular, which means that people can also use the solution without disabilities and the elderly and people with reduced mobility and the mentioned group of users.

3.3 SWOT Analysis for the Development of the InnIoTShop Service

SWOT analysis aims to identify the strengths, weaknesses, opportunities, and threats of the InnIoTShop service. With this analysis, it is possible to determine the advantages and disadvantages and which parts can be improved.

Table 1 shows some advantages, i.e., strengths that can bring positive results for InnIoTSolution and InnIoTStore. The ability to ensure the independence of the user is one of the greatest strengths of the InnIoTShop service, as it raises the user's self-confidence and eliminates the need for other people or aids. A small sample of users dramatically reduces the need for retail chains to implement such a solution because a visually impaired person may be very few in the broader area. The lack of competition could bring an opportunity to present a unique solution to the market that can attract customers. It is essential to ensure the privacy of users, which poses a threat to the proposed solution because it is not possible to say with certainty that an attack on the system will not happen.

Table 1 SWOT analysis of the service

Strengths	Weaknesses
• User independence • Universal design • The solution covers all activities in the purchase process • User mobility • Socially responsible action • Solution modularity	• The implementation of the solution requires the adjustment of the store • Unforeseen errors in the operation of the solution • Limitations of existing technologies • Cost-effectiveness • Application update • A small user sample
Opportunities	**Threats**
• The development of technology makes it possible to improve the solution • By implementing solutions retail chain shows customer care • Improving the quality of life • Lack of competition • Using mobile devices in everyday life • Education of users makes it easier to use solutions	• Large initial investment in the implementation of the solution • Consumer distrust of new technologies • System reliability • User privacy • Lack of investment

4 Ecosystem and Business Model

Figure 4 shows the ecosystem model of the InnIoTShop service that is focused on the user's mobile application and is created in the transformation accelerator tool. It is important to emphasize that Fig. 4 does not show the whole possible solution and does not show the data flows.

At the heart of the ecosystem is the InnIoTShop service. All predefined stakeholders participate in the value chain of service delivery. For a more detailed presentation of the model, stakeholders not previously mentioned in the chapter have been added: investors (required in the initial phase of service development), playIoT (online application store), POS service, and bank (required when paying in InnIoTStore).

InnIoTSolution develops the InnIoTShop service and offers it to InnIoTStore. Investors invest in the development of InnIoTShop at an early stage of service design. The proposed solution involves placing a mobile application on the playIoT store to download the application to their mobile device. InnIoTSolution notifies the support service of all changes related to the InnIoTShop service, which is a separate part of InnIoTSolution, responsible for reporting changes to applications and difficulties related to the use of the service and notifying the factory in charge of manufacturing carts of sensor problems and interruptions. It is available at all times to solution users and InnIoTShop. Equipment suppliers supply the necessary equipment (sensors, actuators, etc.) for smart carts to perform the service on time. The network operator provides the necessary telecommunications infrastructure for the InnIoTShop. Once the user has collected the desired products, payment must be made. In doing so, InnIoTShop sends a payment request to the payment service provider, who then responds to the request by approving or rejecting it.

In the ecosystem designer module within the transformation accelerator tool used to model the ecosystem, it is possible to show the relationship between the stakeholders of the business scenario from five points of view: products/services, contractual, financial, operational, and data. In Fig. 4, they are marked in different colors so that green represents the financial relationship, orange represents the data being exchanged, purple represents the products/services, blue indicates the contractual relationship, while pink represents the operational relationship between stakeholders.

The business model (Fig. 5) must provide business benefits among all stakeholders in a predefined value chain of service delivery. The elements that influence the development of the business model are:

- Key partners
- Key activities
- Key resources
- Value propositions
- Relations between stakeholders
- Channels
- User segmentation

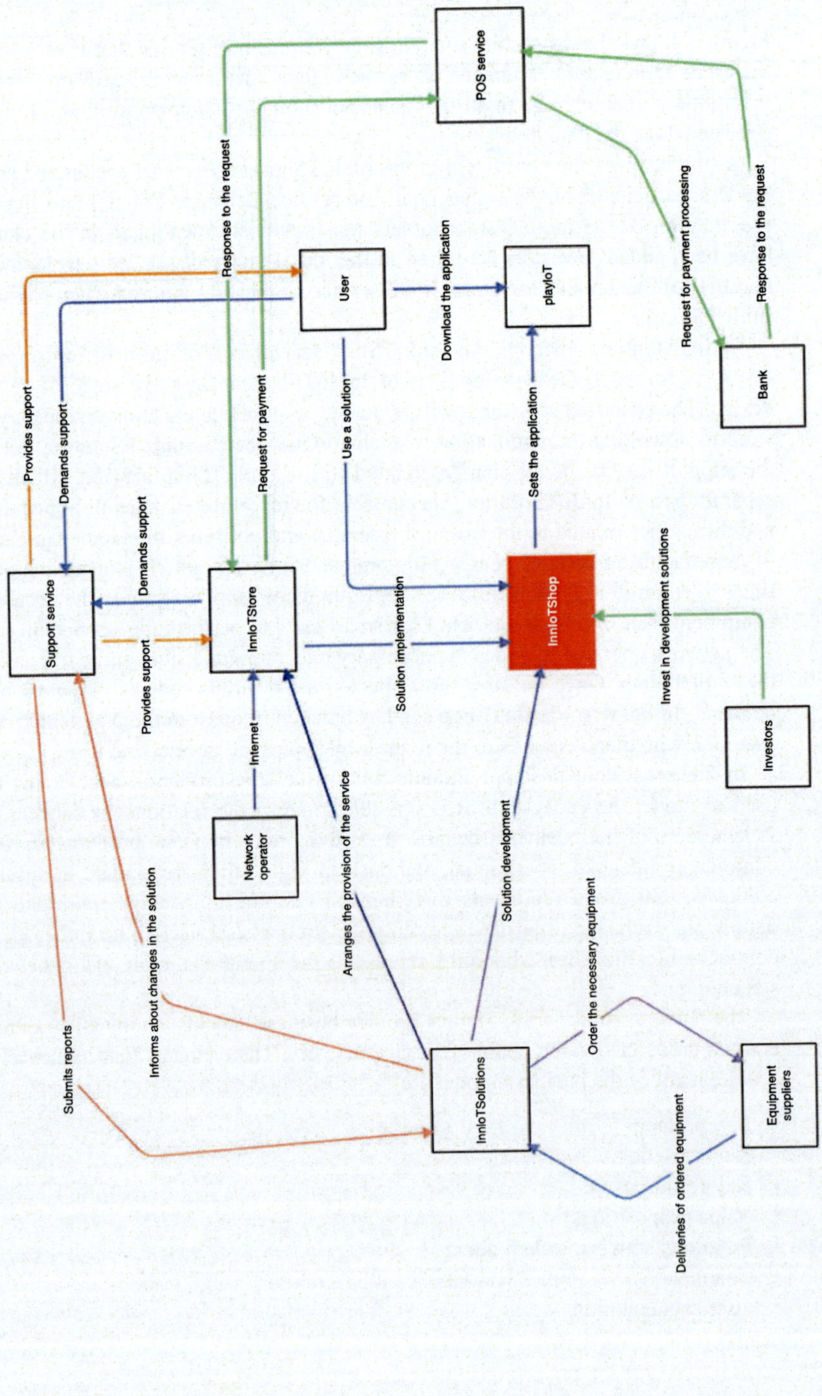

Fig. 4 Part of the elements of the InnIoTShop service ecosystem focused on the user's mobile application

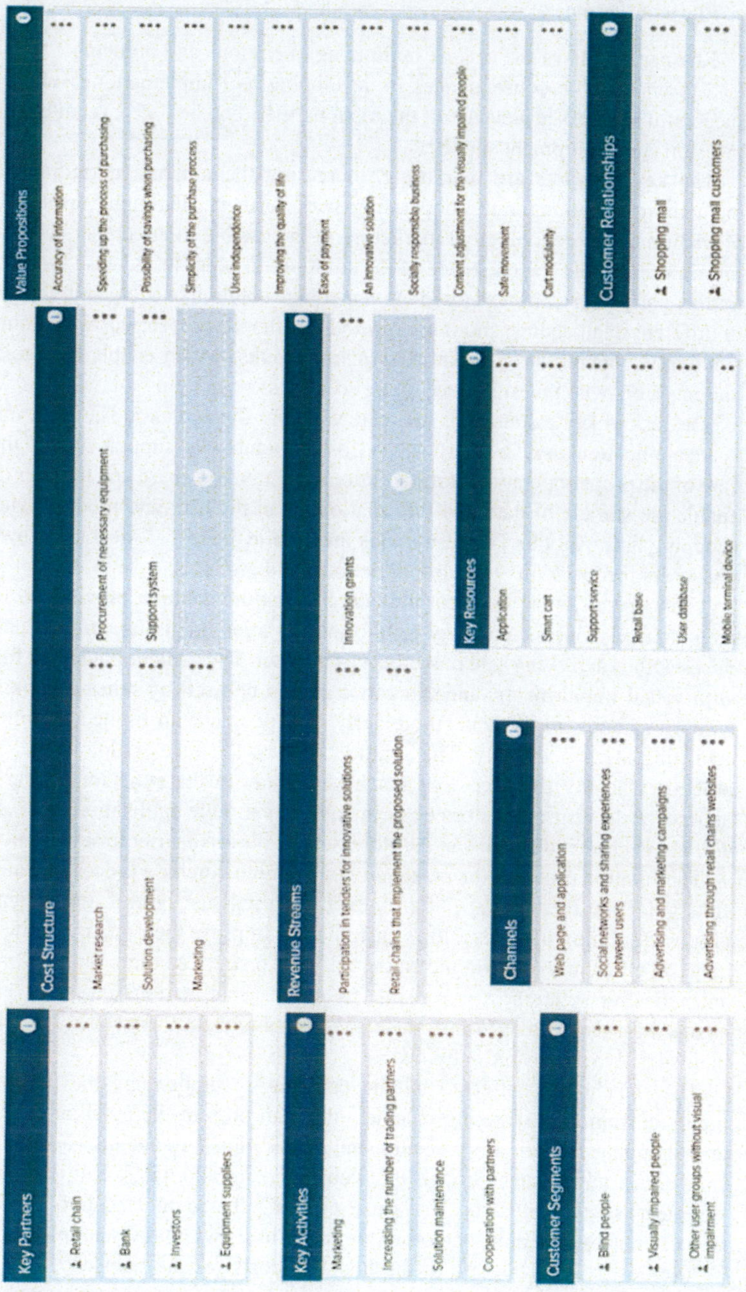

Fig. 5 The business model of the InnIoTShop service

- Cost structure
- Revenue streams

Key partners have the task of optimizing operations and reducing risks that can lead to undesirable consequences in developing and implementing solutions. For successful business achievement, the main partners can be listed: retail chain, bank, investors, and equipment suppliers.

The key activities are marketing, increasing the number of trading partners, maintaining solutions, and cooperating with partners. Through various channels of traditional or technological marketing, it is possible to advertise the solution to inform users about the existence of the solution. In order to increase the number of trading partners, it is necessary to implement the solution in a large number of stores of different retail chains, and it is expected that the support service will maintain the solution. Cooperation with partners such as banks, which enable collecting bills, and investors who invest in developing solutions is essential.

The task of key resources is to create value for the end user. The most important is the application and the smart cart. Also important is the support service that takes care of the proper operation of the solution. The store database and the user database enable the storage of data and information about the user and products and create shopping lists. Another key resource is the mobile device, as most users own it and give users feedback on obstacles, directions, and products.

Value propositions represent the benefits that the solution provides to facilitate the purchase process. Accuracy of information refers to giving correct feedback to the user, thus achieving safe movement of the user. The content is adapted for people with visual impairments, and the cart can also be used by other groups of users (people without impairment, the elderly, and people with limited mobility), thus achieving their modularity. The solution enables the acceleration of the purchase process, whereby the user does not have to stay in the store for a long time or unnecessarily visit other store departments. The mobile application can inform the user about current discounts in the store, which allows the user to save on purchases.

The solution provides the possibility of a simple purchase process because smart carts guide the user to a specific product, cash register, or exit. Implementing the proposed solution in stores allows customers to improve their quality of life because they can efficiently perform the entire purchase process on their own. By presenting an innovative solution, it is possible to attract users, various investors, and retail chains who would like to introduce such a solution in their stores, which proves their socially responsible activities.

The stakeholders affected by the company's solution are the malls and the mall's customers. Marketing channels that can present the solution to customers and shopping centers are websites and applications, social networks, marketing campaigns, advertising through the websites of retail chains, and oral sharing of experiences between users. Users can generally be segmented into blind people, visually impaired people, and other groups of users without visual impairment.

The cost structure refers to the financial costs for the company, and they can be for the areas of market research, procurement of necessary equipment, solution development, support systems, and marketing. Revenue flows indicate the elements based on which the company generates revenue, such as participation in tenders, grants for innovation, and retail chains that implement the proposed solution in their stores.

5 Conclusion

The transformation accelerator tool was used to develop the business model and business scenarios of the InnIoTShop service, and also to apply ecosystem elements that are focused on the user's mobile application. All parts of the value chain of delivery of the InnIoTShop service are connected into one whole. Applying the business model presented in the chapter enables the creation of an efficient management system among all the processes during the delivery of the InnIoTShop service to the end user.

When designing the InnIoTShop service, it is essential to apply universal design principles to allow easy use of the service for all groups of users for whom the service is intended. The development and implementation of the InnIoTShop service can lead to an increase in the quality of life, increase in the degree of independence of people with visual impairments, and increase in the degree of integration of these people into society.

It is essential to continuously improve the proposed concept while monitoring the development trends of modern information and communication technologies. It is also essential to raise awareness of the importance of using assistive technologies while encouraging socially responsible business. Future work will focus on the processes that take place between InnIoTShop service users and smart cart.

References

1. V. Kulyukin, C. Gharpure, J. Nicholson, in *RoboCart: Toward Robot-Assisted Navigation of Grocery Stores by the Visually Impaired*. 2005 IEEE/RSJ International Conference on Intelligent Robots and Systems, (IEEE, 2005), pp. 2845–2850. https://doi.org/10.1109/IROS.2005.1545107
2. V. Kulyukin, C. Gharpure, J. Nicholson, G. Osborne, Robot-assisted wayfinding for the visually impaired in structured indoor environments. Auton. Robot. **21**, 29–41 (2006). https://doi.org/10.1007/s10514-006-7223-8
3. S. Manikandan, C. Arul Murugan, Smart shopping facilitator for visually impaired based on Zigbee. Int. J. Modern Trends Eng. Sci. **3**, 23–24 (2016)
4. M. Periša, V. Anić, I. Badovinac, L. Ćorić, D. Gudiček, I. Ivanagić, L. Terzić, in *Assistive Technologies in Function of Visual Impaired Person Mobility Increases in Smart Shopping Environment*, eds. by L. Knapčíková, D. Peraković, A. Behúnová, M. Periša. 5th EAI International Conference on Management of Manufacturing Systems, (Springer International Publishing, Heidelberg, 2021), p. 350. doi:https://doi.org/10.1007/978-3-030-67241-6
5. A. Yewatkar, F. Inamdar, R. Singh, Ayushya, A. Bandal, Smart cart with automatic billing, product information, product recommendation using RFID & Zigbee with Anti-Theft. Procedia Comput. Sci. **79**, 793–800 (2016). https://doi.org/10.1016/j.procs.2016.03.107

6. Y. Berdaliyev, A.P. James, in *RFID-Cloud Smart Cart System*. 2016 International Conference on Advances in Computing, Communications and Informatics (ICACCI), (IEEE, 2016), pp. 2346–2352. https://doi.org/10.1109/ICACCI.2016.7732405
7. P. Kalaivani, C. Rajan, S. Sandhiya, An automated billing system in shopping malls using bascart. Bull. Sci. Res., 62–68 (2019). https://doi.org/10.34256/bsr1919
8. A. Venugopal, S. Francis, A. Ps, K. Tamang, R. George, Automation of shopping malls using intelligent shopping cart. Int. Res. J. Eng. Technol. **4**, 1712–1714 (2017)
9. M. Elgendy, C. Sik Lanyi, in *Review on Smart Solutions for People with Visual Impairment*. Lecture Notes in Computer Science (including subseries Lecture Notes in Artificial Intelligence and Lecture Notes in Bioinformatics) (2018). pp. 81–84. https://doi.org/10.1007/978-3-319-94277-3_15
10. K. Manjari, M. Verma, G. Singal, A survey on Assistive Technology for visually impaired. Internet Things. **11**, 100188 (2020). https://doi.org/10.1016/j.iot.2020.100188
11. B. Priyalakshmi, S. Murugaveni, S. Umamaheswari, Robokart for visually impaired people. Mater. Today Proc., 2–6 (2021). https://doi.org/10.1016/j.matpr.2021.01.087
12. D. Peraković, M. Periša, I. Cvitić, P. Zorić, Business Process Modeling in Industry 4.0 Using Transformation Accelerator Tool, in *New Approaches in Management of Smart Manufacturing Systems*, ed. by L. Knapcikova, M. Balog, D. Peraković, M. Periša, (Springer, Cham, 2020), pp. 231–248. https://doi.org/10.1007/978-3-030-40176-4_15
13. T. Keiningham, L. Aksoy, H.L. Bruce, F. Cadet, N. Clennell, I.R. Hodgkinson, T. Kearney, Customer experience driven business model innovation. J. Bus. Res. **116**, 431–440 (2020). https://doi.org/10.1016/j.jbusres.2019.08.003
14. M. Geissdoerfer, D. Vladimirova, S. Evans, Sustainable business model innovation: a review. J. Clean. Prod. **198**, 401–416 (2018). https://doi.org/10.1016/j.jclepro.2018.06.240
15. M. Periša, D. Peraković, I. Cvitić, M. Krstić, Innovative ecosystem for informing visual impaired person in smart shopping environment: InnIoTShop. Wirel. Netw **5** (2021). https://doi.org/10.1007/s11276-021-02591-5

Part III
Smart Techniques/Smart Technology

Polyvinyl Butyral (PVB) as a Significant Material Used by Additive Manufacturing

Lucia Knapčíková, Annamária Behúnová, Jakub Kaščak, and Jozef Husár

1 The State of the Art

Since 2000, the most popular method in additive production has been fused deposition modelling (FDM) technology [1]. It uses the direct application of material by melting in a printhead and applying point by point. The most common materials are thermoplastics. The most common areas of application of additive manufacturing are described in Fig. 1.

It is the most widespread form of 3D printing used mostly by open source. Adrian Bowyer, who was able to print a new 3D printer from a 3D printer in 2008 and subsequently released his designs under the name RepRap, had a great deal of credit for expanding this type of printer [2]. According to the Wohlers Report, Stratasys sold almost as many FDM devices in 2003 as all other AV devices with other technologies combined. In 2006, Stratasys sold 54.7% of all AV equipment sold. However, few people know that 2007 was a turning point for FDM technology [3]. This year, the RepRap project focused on FDM technologies was launched based on

L. Knapčíková (✉) · J. Husár
Department of Industrial Engineering and Informatics, Faculty of Manufacturing Technologies with a Seat in Prešov, The Technical University of Košice, Prešov, Slovak Republic
e-mail: lucia.knapcikova@tuke.sk; jozef.husar@tuke.sk

A. Behúnová
Faculty of Mining, Ecology, Process Control and Geotechnologies, Institute of Earth Resources, Technical University of Košice, Košice, Slovak Republic
e-mail: annamaria.behunova@tuke.sk

J. Kaščak
Department of Computer Aided Manufacturing Technology, Faculty of Manufacturing Technologies with a Seat in Prešov, The Technical University of Košice, Prešov, Slovak Republic
e-mail: jakub.kascak@tuke.sk

L. Knapčíková, D. Peraković (eds.), *6th EAI International Conference on Management of Manufacturing Systems*, EAI/Springer Innovations in Communication and Computing, https://doi.org/10.1007/978-3-030-96314-9_9

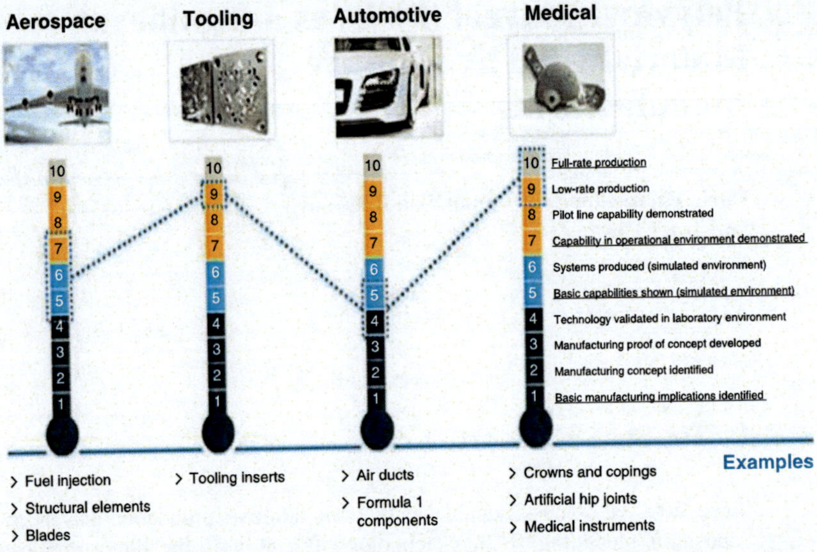

Fig. 1 Possibilities of using additive manufacturing. Introduction to Additive Manufacturing Technology, EPMA, 2nd Edition, 2017 (Roland Berger) [1]

open solutions. Thanks to this, according to the Wohlers Report 2016 study, in 2015 we managed to sell almost 12,600 AV devices with a price below 5000 USD. The RepRap project became the basis for the emergence of hundreds of new startups that began producing and selling PDM screens worldwide [4].

By using additive manufacturing in practice, we achieve that printing begins on a clean substrate by gradually applying the material in layers as accurately as possible [2, 5]. In this way, the printer transfers printed material in parts of the v2D space. The next layer is already moving horizontally until a complete 3D model of the product is created. We can say that the particular methods of layer printing differ, but the material is common [3]. The thickness of the layers best corresponds to the quality of the model (Fig. 2).

Therefore, the smallest possible layers' process seems to be of the most optimal quality, but it has its limitations. Hardware limitations are in the formation of layer thicknesses [6]. Although it is technologically possible, we spend a lot of time with software support. The printing time itself is significantly extended, as it takes longer to create one layer and at the same time has to make a larger number of layers. Polyvinyl butyral has similar printing properties as PLA, and mechanical properties identical to CPE or PETG [4, 6].

Thanks to the clarity and the possibility of easy smoothing, it is optimal, especially for visual models, such as nontraditional vases, jewelry, lights, and other design elements. Models that need to use the clarity of a filament seem to be completely ideal. Here it is good to play with different printing settings, such as layer thickness, perimeters, and filling [2, 5]. PVB is not suitable for printing

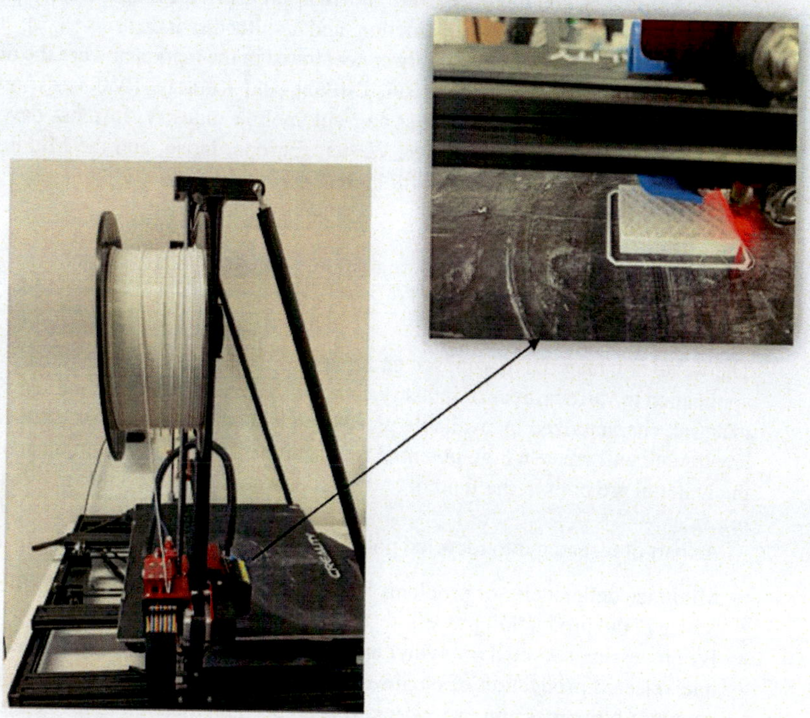

Fig. 2 The Creality 3D printer used by our research with detail on printed material. [Authors' own processing]

technical models. Although its spatial stability surpasses most filaments and its toughness is close to materials such as CPE (and more brittle PETG), its temperature resistance, tensile strength, and cohesiveness of the layers are more like PLA.

It is important to note that PVB is not the most suitable for printing mechanical parts. It is similar to CPE (PETG) due to its dimensional stability, tensile strength, and impact strength. However, its thermal efficiency and adhesion between the layers are identical to PLA [6, 7]. In terms of PVB production, several companies are currently based in Europe and around the world. A substantial part of the use of PVB is focused on the production of films. In general, we can say that PVB resins are used to produce protective layers, adhesives as a basis for the production of dry paints, and an additive for ceramics and composite fibers. PVB film is characterized by properties such as high tensile strength, impact resistance, transparency, and flexibility. These properties are essential in the production of safe glass. It is precise because of the content of alcohol, ester, and acetate bonds that the PVB film holds the glass firmly, even in the event of an impact [7]. The glass adheres to the intermediate layers of the PVB film and thus prevents breakage. Safety glass is essential for the automotive, construction, and architecture industries [8, 9].

Polyvinyl butyral resins are highly concentrated in the market and are the domain of four companies—Astman, Sekisui, DuPont, and Kuraray. PVB is exported to countries with an expansion of a strong automobile industry. In most developed countries, such as the United States, Western Europe, Japan, and the Middle East, growth in the demand for PVB is still high [9].

1.1 The Aim of the Research

The waste raw material of windscreen films thus represents new possibilities for its application to various areas of industry. Due to the already defined advantages of this material, characterized by available processing technologies and low input costs, it creates and will create a high potential for usability. The development tendencies of the material are both in the field of study and testing and their direct application in practice.

As part of research into recycled polyvinyl butyral, we plan for the following:

- Minimize deficiencies or problems with waste material from car windshields.
- Reuse in the production process.
- By processing recycled polyvinyl butyral into new materials to implement them into selected production of engineering products and not only there. Due to its nature of properties and processing, it is an open application to the construction industry, garden construction, and road construction.
- Use the materials produced based on recycled polyvinyl butyral to save the company's initial investment.
- Products based on recycled polyvinyl butyral should be friendly to the environment and human health.
- Recyclable.

2 Polyvinyl Butyral (PVB) as a Key Material

Polyvinyl butyral is an important part of car windows or building windows. Through recovery technology, mainly material recovery takes the form of flakes with dimensions of 2–20 mm and a thickness of 0.5–1.5 mm. The polyvinyl butyral recycled [10] in this way was contaminated (dust, glass fragments), so it was necessary to wash and dry the material before starting laboratory work thoroughly. In our research PVB was used in the granulate form (Fig. 3).

Table 1 describes the basic properties of recycled polyvinyl butyral.

From the chemical perspective, polyvinyl butyral is a thermoplastic material soluble in ethanol, butanol, ethyl acetate, and butyl acetate in mixtures of chlorinated

Fig. 3 Recycled polyvinyl butyral (left—granulate, right—homogenized material). [Authors' own processing]

Table 1 Polyvinyl butyral properties [1]

Recycled polyvinyl butyral	
Shape	Granulate
Color	White/gray
Size	20–30 mm
Purity	More than 97%
Impurity content	Less than 3%
Residual humidity	ca. 2%
Proportion of glass content particles	Less than 2%

hydrocarbons insoluble in aliphatic hydrocarbons (in gasoline). The density of the polyvinyl butyral used in the research was 1.07 g cm^{-3}. The price of the recyclate on the plastics market is 0.25–0.50 €/kg [11].

2.1 *Differential Scanning Calorimetry (DSC) Analysis*

Differential scanning calorimetry (DSC), or differential compensation calorimetry, is a method in which the thermal properties of materials are examined. This method is widely used to determine melting points, glass transitions, and crystallization of various materials [12]. It finds specific applications in the study of polymers, emulsions, reaction kinetics, or thermal conductivity of materials. The standard temperature range of the instruments is −100 to 600 °C. DSC method of measurement consists of a constant heating (cooling) rate of two vessels; one vessel is empty (reference), and the other contains the sample [13]. In this case, the control unit (a computer) constantly ensures a constant heating rate of both samples throughout the experiment, typically 10 °C/min. Since one vessel is empty and the other contains the sample, the heat flux to the individual vessels will vary due to the composition of the sample and the phase changes taking place in the sample. In differential scanning calorimetry, we focus on determining the glass transition and melting point [14]. The glass transition is associated with a change in the internal structure of the materials, which is reflected, in particular, in the mechanical properties. Below the glass transition temperature, the material is brittle and hard. Above the glass transition temperature, the material becomes flexible [15].

As a rule, the glass transition does change the mechanical properties of the material and the heat capacity, which we use in the measurement. The glass transition temperature (Tg) is the value in the middle of this transition, i.e., the curve's inflexion point. Material melting is an endothermic process [16]. The material temperature remains constant despite continuous heating, so heat is absorbed into the sample, and this energy is converted to melting [17]. This phenomenon manifests in the same peak shape on the DSC curve as the crystallization but opposite. The melting point (Tm) is declared as the peak, and the energy consumed for melting is obtained by integrating the height [18].

3 Results and Discussion

The glass transition temperatures (Tg) and melting point (Tm) temperatures for recycled polyvinyl butyral build important information obtained after differential scanning calorimetry. We receive information about the processing possibilities of the test material [19]. The method used in the testing is called differential scanning calorimetry (DSC). It is also used to quantify enthalpy characteristics associated with substances' physical and chemical transformations in test systems [20]. Using DSC, we can measure heat capacity and determine the purity of substances. The material was analyzed under laboratory conditions, at a room temperature of 22 °C and a humidity of 60%, following DIN ISO 113 57.

The DSC 204 instrument was used for analysis. The NETZSCH Proteus program evaluated NETZSCH (Fig. 4) and the obtained data.

In DSC analysis, 9 mg sample was used, with the temperature from −50 °C to 300 °C. The measurement process itself began by regress of the temperature to −180 °C and then heating it. The key temperature to start the analysis was −50 °C. The piece is heated only after reaching a temperature of +300 °C (Figs. 5 and 6) by the method of differential scanning calorimetry. The sample is subjected to linear

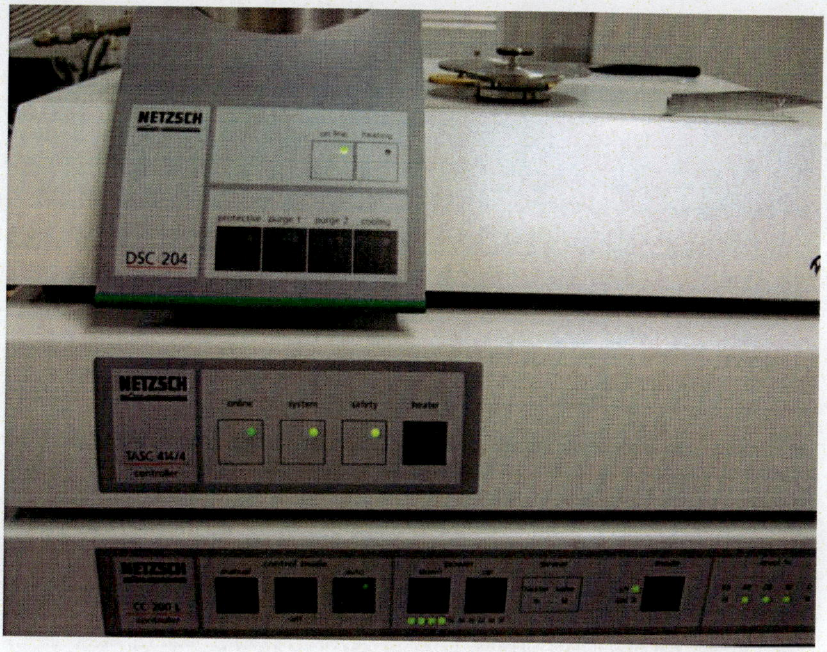

Fig. 4 Differential scanning calorimetry equipment NETZSCH. [Authors' own processing]

heating and thus to the sample's heat flow rate [21]. The heat flux is proportional to the instantaneous specific heat. Two symmetrical vessels were stored inside the measuring device, with the internal temperature equal to the room temperature, i.e., ca. 20 °C. The instrument includes a resistance thermometer and a heating element built into the sample carrier and serves as the primary control of the system. The following graphs represent the data obtained after DSC analysis of recycled polyvinyl butyral samples.

The temperature of the sample is maintained isothermally with the reference. Such an amount of energy is needed to maintain isothermal conditions [22]. Generally, material placed on metal pads reduces the thermal gradient to a minimum [23, 24]. When using DSC, high heating rates are used (in tens of K min^{-1}, °C min^{-1}). In this way, the required temperature range is guaranteed.

The difference between the first and second heating can be seen from the NETZSCH Proteus evaluation program [18]. It means that during the first heating, e.g., there is inaccurate seating of the sample in the test aluminum container. In the second case of heating, a final analysis of the material is obtained. Upon completion of the examination, the values for the glass transition temperatures are from 62 to 72 °C. The melting point Tm is higher, at about 150 °C. At 15 °C, temperature changes in the material begin, but without disturbing the internal structure. The temperature values of 217 °C and 253 °C show the disruption of the internal

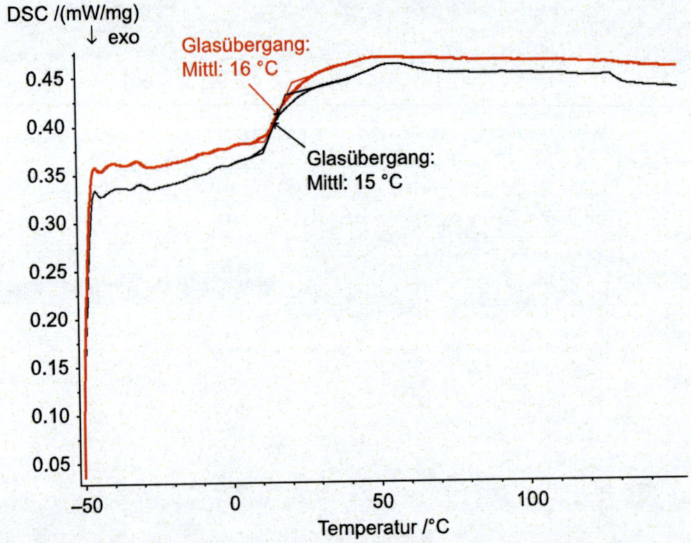

Fig. 5 Differential scanning calorimetry for recycled PVB. [Authors' own processing]. German language "Glasübergangstemperatur": Glass transition temperature (Tg); German language "Temperatur": Temperature; Red curve: Recycled polyvinyl butyral (1. heating); Black curve: Recycled polyvinyl butyral (2. heating)

bonds of the material. The results show that the analysis was performed correctly, according to the relevant standard DIN ISO 113 57.

4 Conclusion

The competition and competitive position of recycled polyvinyl butyral materials and their subsequent application to selected components of engineering production represent the quality requirements on the European market and respect the legislative measures of the country in which the product is manufactured. Additive manufacturing is one of the basic supporting technologies that shape new approaches to manufacturing and the products and plants of the future. With this revolution in manufacturing, the digital process will enable Europe to return production from lower wage areas to encourage innovation and create sustainable growth at home. At present and in the future, not excluding, the priority will be respect and incorporation into product designs of customer requirements, the so-called production customization.

The use of recycled materials in additive production is conditioned mainly by the qualitative properties of the input material. Our priority is to work with plastic waste, which we recycle using appropriate technology.

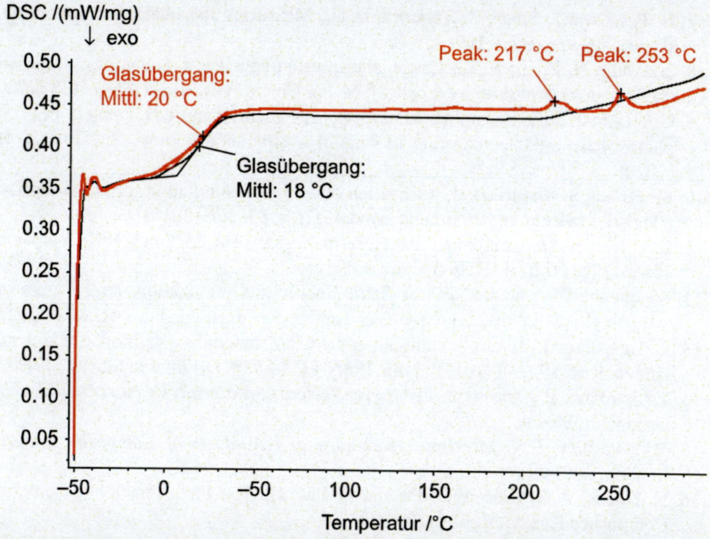

Fig. 6 Recycled polyvinyl butyral (left—granulate, right—homogenized). [Authors' own processing]. German language "Glasübergangstemperatur": Glass transition temperature (Tg); German language "Temperatur": Temperature; Red curve: Recycled polyvinyl butyral (1. heating); Black curve: Recycled polyvinyl butyral (2. heating)

Plastic flocks or granulates. It will then process the 3D printer, and plastic waste will help improve public spaces' quality according to the circular economy's ideology.

Greater design freedom leads to new products: almost unlimited design freedom can bring a wide range of benefits in different areas (i.e., automotive and aerospace, healthcare, machinery and equipment, sports equipment, and lifestyle).

References

1. Polyvinyl butyral. https://www.ihs.com/products/polyvinyl-butyral-chemical-economics-handbook.html. Accessed 21 Nov 2016
2. New trends by additive manufacturing. https://www.atpjournal.sk/novetrendy/aditivna-vyroba-aktualny-stav-a-nove-trendy.html?page_id=24729. Original in Slovak. Accessed 14 Apr 2021
3. A.K. Dhaliwal, J.N. Hay, The characterization of polyvinyl butyral by thermal analysis. Thermochim. Acta **91**, 245–255 (2002). https://doi.org/10.1016/s0040-6031(02)00187-9
4. Polyvinyl butyral. https://www.ihs.com/products/polyvinyl-butyral-chemical-economics-handbook.html. Accessed 10 Apr 2021
5. C.M.F. Barry, S.A. Orroth, Processing of Thermoplastics, in *Modern Plastics Handbook*, ISBN 0-07-026714-6, ed. by C. A. Harper, (2000)
6. Kuraray.eu. www.Kuraray.eu. Accessed 21 Mar 2021

7. H. Pacaiova, J. Sinay, R. Turisova, et al., Measuring the qualitative factors on copper wire surface. Measurement **109**, 359–365 (2017)
8. L. Wang, H. Zhou, N. Li, Carrier transport composites with suppressed glass-transition for stable planar perovskite solar cells. J. Mater. Chem. A **8**(28), 14106–14113 (2020)
9. G. Heath, T. Silvermann, M. Kempe, et al., Research and development priorities for silicon photovoltaic module recycling to support a circular economy. Nat. Energy **5**(7), 502–510 (2020)
10. M. Straka, S. Khouri, et al., Utilization of computer simulation for waste separation design as a logistics system. Int. J. Simulat. Model. **17**(4), 83–596 (2018)
11. C.A. Harper, *Handbook of Plastics Technologies* (The McGraw-Hill Companies, 2006), pp. 15–35, ISBN 0-07-146068-3
12. Parameters for Standard Linear Static Topology Optimization. https://abaqus-docs.mit.edu/2017/English/TsoUserMap/tso-c-user-TopOpt-Sett-LStaTop.htm. Accessed 21 Dec 2020
13. L. Knapčíková, Životný cyklus kompozitného materiálu s termoplastickou matricou. Strojárstvo. Roč. **19**, 7–8 (2015)., s. 83. ISSN 1335-2938. Original in Slovak (2015)
14. T. Liptáková, P. Alexy, et al., *Polymérne Technické Materiály* (Vysokoškolská učebnica, 2009)., Original in Slovak
15. V.V. Vasiliev, E.V. Morozov, *Mechanics and Analysis of Composite Materials* (Elsevier, London, 2001)
16. N. Dodds, A.G. Gibson, D. Dewhurst, J.M. Davies, Fire behavior of composite laminates. Composites Part A. **31**, 689 (2000)
17. L. Knapčíková, M. Herzog, P. Oravec, *Material Characterization of Composite Materials from Used Tires*, vol 4 (Výrobné inžinierstvo, 2010), pp. 31–34. Original in Slovak
18. J. Liu, A.T. Gaynor, S. Chen, Z. Kang, K. Suresh, A. Takezawa, L. Li, J. Kato, J. Tang, C.C.L. Wang, et al., Current and future trends in topology optimization for additive manufacturing. Struct. Multidiscip. Optim. **57**, 2457–2483 (2018)
19. M. Straka, A. Rosova, et al., Principles of computer simulation design for the needs of improvement of the raw materials combined transport system. Acta Montan. Slovaca **23**, 163–174 (2018)
20. A.P. Mouritz, Z. Mathys, Mechanical properties of fire-damaged glass-reinforced phenolic composites. Fire Mater. **24**, 67–75 (2000)
21. S. Fiebig, J. Sellschopp, H. Manz, T. Vietor, K. Axmann, A. Schumacher, in *Future Challenges for Topology Optimization for the Usage in Automotive Lightweight Design Technologies*. Proceedings of the 11th World Congress on Structural and Multidisciplinary Optimization, Sydney, Australia, 7–12 June 2015, vol 142, pp. 1–8
22. R. Singh, J.P. Davim, *Additive Manufacturing: Applications and Innovations*, 1st edn. (CRC Press, Boca Raton, FL, 2018), p. 280
23. N. Li, S. Huang, G. Zhang, R. Qin, W. Liu, H. Xiong, G. Shi, J. Blackburn, Progress in additive manufacturing on new materials: a review. J. Mater. Sci. Technol. **35**, 242–269 (2019)
24. J. Sinay, M. Balazikova, M. Dulebova, et al., Measurement of low-frequency noise during CNC machining and its assessment. Measurement **119**, 190–195 (2018)

Optimization of the Production Process and Defect Identification in the Production of Face Shields Using Fused Deposition Modeling Technology

Jozef Husár, Lucia Knapčíková, and Jakub Kaščak

1 Introduction

The term 3D printing, i.e., additive manufacturing, encompasses a variety of processes and technologies that offer solutions for producing parts and products created from numerous materials. All these processes and technologies have in common the principle that the production method is carried out in individual layers. Additive technologies are contrast against manufacturing technologies such as turning and milling.

Today, this technology is described as a revolutionary production of products enabling their unconventional development regardless of the subsequent method of production with the minimization of waste material. This should bring a whole new perspective on industrial production in the future, as we know it today.

We find the application of 3D printing in sectors such as medicine, industry, art, design, and also education. The main advantage of this technology is considered to be the ability to adapt the product to customer requirements. 3D printing is notable for its contribution to prototyping and product development.

The aim of this chapter is to describe 3D printing technologies with a focus on the most used 3D printing—FDM technology, which we applied in the first wave of the COVID pandemic to the printing of protective shield frames. We will gradually

J. Husár (✉) · L. Knapčíková
Department of Industrial Engineering and Informatics, Faculty of Manufacturing Technologies with a Seat in Prešov, The Technical University of Košice, Prešov, Slovak Republic
e-mail: jozef.husar@tuke.sk; lucia.knapcikova@tuke.sk

J. Kaščak
Department of Computer Aided Manufacturing Technology, Faculty of Manufacturing Technologies with a Seat in Prešov, The Technical University of Košice, Prešov, Slovak Republic
e-mail: jakub.kascak@tuke.sk

L. Knapčíková, D. Peraković (eds.), *6th EAI International Conference on Management of Manufacturing Systems*, EAI/Springer Innovations in Communication and Computing, https://doi.org/10.1007/978-3-030-96314-9_10

describe the process of optimizing the production process, adjusting the parameters and defects that occurred during 3D printing [1].

1.1 Additive Production

Additive production, or 3D printing, is characterized as a production process in which the product is created by gradual application of material in individual layers, which are connected to each other, for example by melting or gluing. In contrast to conventional production methods, such as subtractive machining, 3D printing can also produce very complicated shapes and process difficult-to-machine materials. 3D printing in the field of prototyping (rapid prototyping) or in small series production has become an unrepeatable tool. All commercial devices of additive technologies use the principle of layering the material. However, they differ in the way the layers are formed and interconnected. The various technologies then indicate the accuracy of the object produced, mechanical properties, and production speed [2, 3].

1.2 History of Additive Production

The history of 3D printing began to be written only in the second half of the twentieth century, specifically in 1986, when Charles Hull patented stereolithography technology. It is a method used to create objects by sequentially curing polymers [4].

In the late 1990s, Charles Hull created the first device for the general public that could print in 3D. It was a stereolithographic apparatus SLA-1, which was a form of 3D printer, although it was not referred to as a 3D printer at that time. However, SLA models from that time became the basis for the development of today's 3D printers.

For a long time, 3D Systems, founded by Charles Hull, held the leading position in the 3D printing market. However, with the advent of competition, new modern technologies began to emerge, which still have their place in 3D printing. While Kodama, Le Mehaute, and Hull worked on stereolithography technology, others developed sintering and extrusion techniques for additive layering of objects [5].

Additional 3D printing technologies were released in the following years, including FDM from Stratasys and SLS from 3D Systems. FDM technology consists of melting a polymer fiber, which is applied in layers. This is the simplest and most common of these technologies. These three technologies—SLA, SLS, and FDM—have remained the dominant additive manufacturing techniques and each has its strengths.

Open-source project RepRap, founded in 2004 by Adrian Bowyer, is based on the idea of printing spare parts, which significantly accelerated the development of a 3D printer. In 2009, patents for FDM technology expired, which enabled a sharp drop in prices and thus a mass expansion of this technology for end users [6].

1.3 Types of Additive Production

At present, there is a wide range of additive technologies; their differences are mainly in the way in which the individual layers of the model are applied, and range of usable materials used for object creation. Choosing the most suitable additive production process for a particular application is very difficult (Fig. 1).

For our needs, the material extrusion method was chosen, specifically FDM: FDM (FFF)—fused deposition modeling/fused filament fabrication.

This technology uses direct application of the material by melting in the printhead; after extruding it reaches the required temperature and afterwards it creates desired model layer by layer. The most common materials are thermoplastics. FDM technology is also the most widespread form of 3D printing; at the same time it is the least economically demanding 3D printing technology. The disadvantages of this technology are the long printing time and low print quality [7] (Fig. 2).

The first step in 3D printing is to convert the CAD format (.stl, .obj, or other) to the g-code format (.gcode) required for the 3D printer. The program contains the coordinates according to which the nozzle moves, print and extrusion speed, and other operat ing parameters required for successful model creation.

FDM printing is based on the principle of melting a material that is in the form of a plastic string, a filament. The filament is wound on a spool from which it is unwound as required. The printhead then prints the physical 3D model layer by layer according to the digital model. The molten material is extruded through a nozzle onto a production platform where it is cooled by active fans and after the subsequent solidification, the final model is created [8, 9].

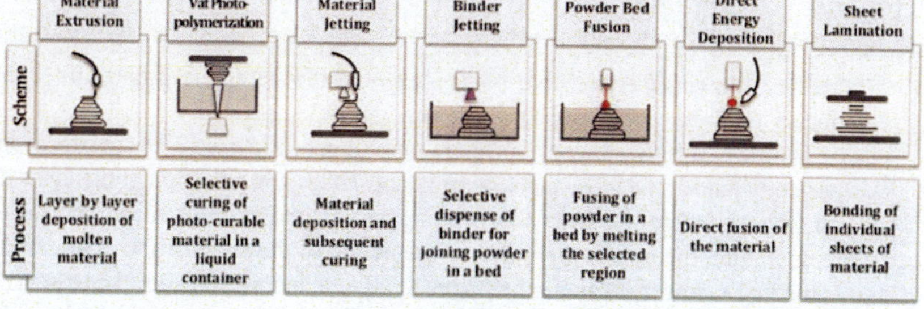

Fig. 1 3D printing methods

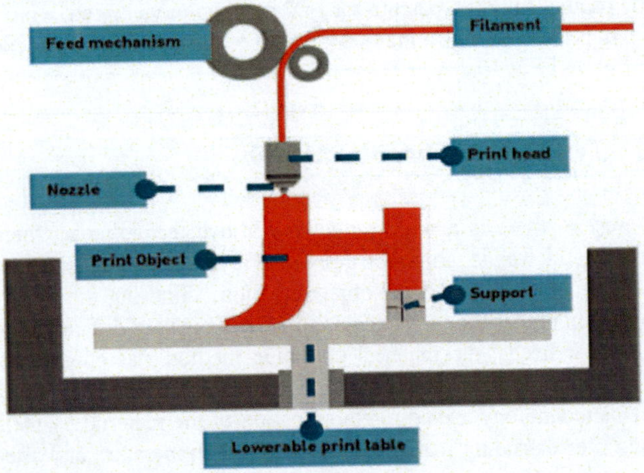

Fig. 2 Principle of FDM printing

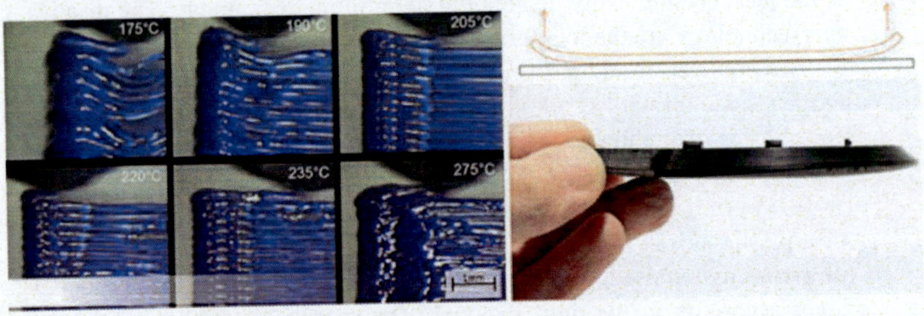

Fig. 3 Layering at different temperatures and poor adhesion of the first layer

1.4 Parameters and Causes of Printing Errors Using FDM Technology

In the following section, we have presented the parameters that affect the print quality on the example of FDM printing.

Printing surface, layer adhesion, and layer thickness—the surface must be perfectly flat for the most accurate prints possible. The adhesion of the layer depends on several factors such as the printing speed, the type of material, the printing temperature, the nozzle diameter, or the height of the layer. Build plate temperature is the first factor that causes the platen to adhere and prevent shrinkage [10] (Fig. 3).

Layer thickness: With FDM printing it is possible to set different layer heights; the layer height of 50–800 μm is used (Fig. 4).

Nozzle diameter: This is variable from 0.2 to 3 mm based on the used filament diameter. The diameter of the nozzle directly affects the width of the extrusion.

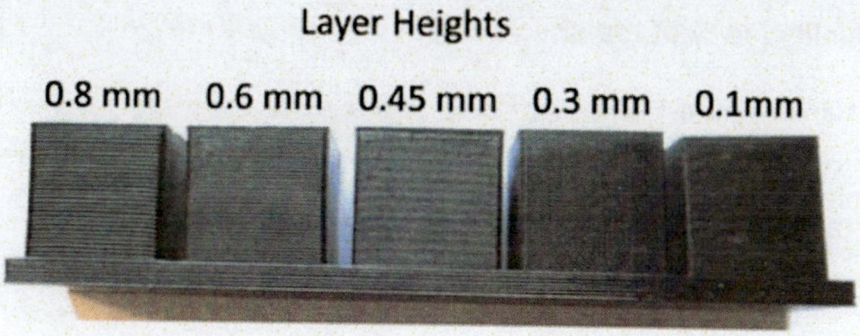

Fig. 4 Layer height in the range of 0.1–0.8 mm

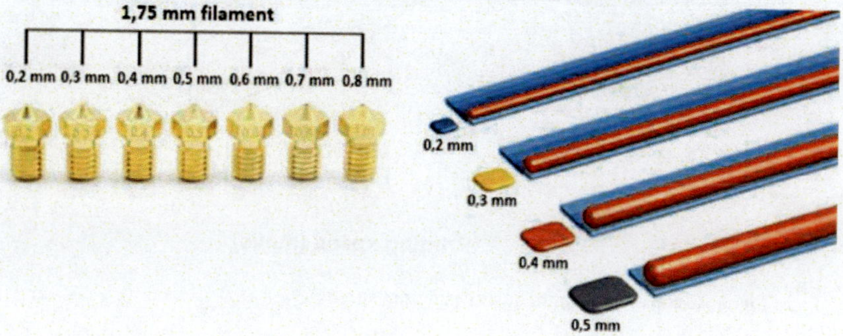

Fig. 5 Nozzle size and parameter comparison

While printing, it is necessary to monitor that the extruder dispenses the correct amount of filament. This process has significant meaning for dimension accuracy of the created model and print quality. When choosing the size of the nozzle diameter in a 3D printer, it is necessary to consider what is important and which elements need to be emphasized (Fig. 5).

Print speed: Visually, a slower print speed supports a higher quality result, although the time saved at higher speeds may be authoritative for some manufacturers. It is safe to say that lower print speeds are required for high-quality production [10] (Fig. 6).

Nozzle temperature: This affects the visual results, and the mechanical properties of the print. The type of material used is important (Fig. 7).

Infill pattern and infill percentage: The infill percentage of the model played an important role in this type of design. The design of the face shields shows that it is a relatively thin-walled design. As a result, there is a risk of deformation of the outer perimeter walls caused by the infiltration beyond the outer perimeter. This error is caused by incorrect selection of parameters when printing thin-walled components. When the infill pattern was transmitted outside the outer perimeter, it caused a deformation of the peripheral part of the model. With increasing diameter of the used nozzle, this shortcoming became more pronounced. Solution to this situation was an excessive requirement for the number of outer perimeters; the device thus

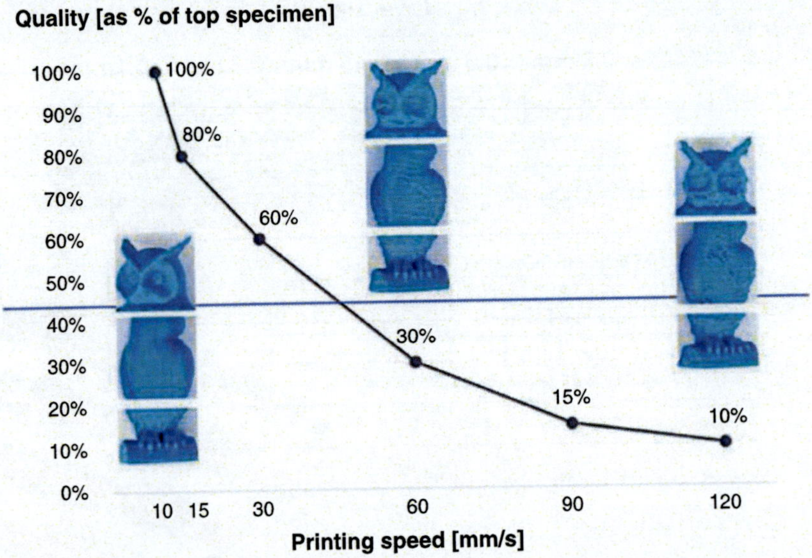

Fig. 6 Comparison of print quality and print speed

Fig. 7 Comparison of print quality and temperature

created is a model made only of circumferential perimeters with a 20% overlap. This option proved to be the most suitable in terms of the production time and durability of face shields [11].

Number of outline perimeters, top bottom layers: As mentioned, the face shield model required delicate adjustment in the number of circumferential perimeters and the infill rate. Closing the model with top and bottom layers was necessary in terms of its durability and quality of the final design. The adjustment of model is necessary due to the formation of elephant's foot during the pressure of printing on the initialization layer. Due to its shape and sharp edge, this could interfere with long-term wear or cause injury. At the same time, it was necessary to ensure sufficient pressure and adhesion of printed model to the build plate.

1.5 COVID-19

SARS-CoV-2-COVID-19 is a severe acute respiratory syndrome. Coronavirus-2 (SARS-CoV-2) is the name given to the new coronavirus in 2019. COVID-19 is the name of a disease associated with the virus. SARS-CoV-2 is a new strain of coronavirus that has not been previously identified in humans. Coronaviruses are viruses that circulate among animals. Some coronaviruses can infect humans. The new coronavirus detected in China in 2019 is closely related to the SARS-CoV-1 virus. SARS appeared in China at the end of 2002 and caused more than 8000 cases in 33 countries in 8 months. About one in ten people who developed SARS died. In the first 11 months of the COVID-19 pandemic (31 December 2019 to 14 December 2020), there were more than 71 million cases and more than 1,600,000 deaths worldwide. EU/European Economic Area data (from countries with available data) show that up to 30% of diagnosed cases of COVID-19 are hospitalized depending on the age and up to 20% of hospitalized people need ventilation support. However, it is important to note that people with more severe symptoms are more likely to be tested than people with less severe symptoms. The actual proportion of people in need of hospitalization out of the total number of infected people is therefore lower than these figures suggest. Hospitalization rates are higher in people aged 60 and over and in people with basic health problems [12].

On the basis of the above, the European Center for Disease Prevention and Control has established basic preventive recommendations in the fight against COVID-19:

- Do not touch your face: The virus can enter your body through your eyes, nose, and/or mouth, so it is important that you do not touch your face with unwashed hands.
- Wash your hands: Frequent washing of hands with soap and water for at least 20 s or thorough cleaning of hands with alcohol-based solutions, gels, or tissues is recommended in all settings.
- Stay away from infected people: Avoid close contact with people infected with COVID-19.

- Avoid social gatherings: Avoid physical gatherings, events, and other social gatherings in areas of ongoing community transmission, and follow local recommendations for mass gatherings.
- Use the face mask: Wear the face mask indoors and outdoors whenever a physical distance from other people cannot be guaranteed.

2 Methodology

Due to the already mentioned factors, two types of 3D printers were used for the printing of shaped shields, namely Creality Ender 3 Pro and Creality CR 10 max. These are simple FDM printers that are affordable. PETG Plasty Mladeč filament was used as the material for the printing of shaped shields. It is a strong and tough material, which is characterized by good temperature stability. This material is suitable for printing large objects due to its small shrinkage during printing [13]. Unlike other materials, PETG is more resistant to acids and solvents, with high and low temperatures. It is possible to thermoform an object extruded from it or to polish it with heat. This material was chosen because of its properties and especially its relative flexibility and resistance to higher temperatures. Face shields used by healthcare professionals and first-contact people are often used for several hours. Therefore, they must be disinfected or heat sterilized to prevent the transmission of viruses (Fig. 8).

Fig. 8 3D printers used to print shape shields

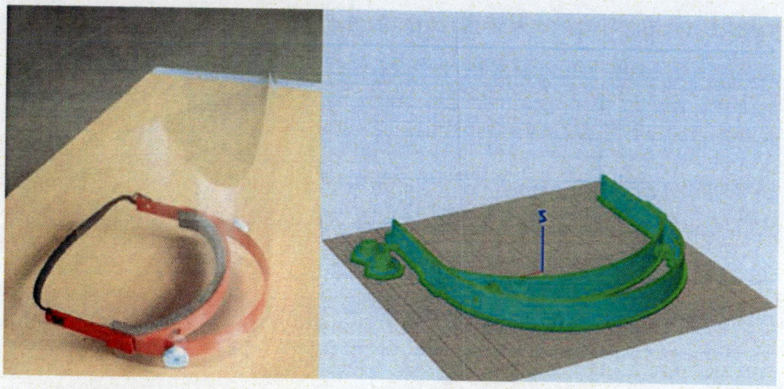

Fig. 9 The final version of the shape shield

As mentioned in the introduction, our aim was to choose among the large number of 3D models of face shields that best meet the conditions of rapid production. The design of face shields consisted of a protective frame, protective foil, reinforcement, 2× nut and screws, and 1× cover for nut. As part of the preparation of the g-code, all the components necessary for the production of one face shield were placed on one printing pad. Thus, the optimization process was implemented for three components. As for the Creality CR max device, and the storage of more components on the build plate, there was logically an increase in the time required for production. This was caused by printhead transitions when printing other components [14] (Fig. 9).

During the process of printing face shields, we encountered a number of problems, especially with the adhesion of the surface when there was shrinkage of the material and subsequent detachment. Also for the first pieces, the problem was a long printing time. We tried to remove this by changing the initial layer speed and the initial layer height. Proper way of cooling, or rather its absence, also played one of the key tasks in this case.

Table 1 shows us the time required for the production of individual components and the printing parameters used. As mentioned, the process is optimized for the production of one complete face shield, which consists of three components [15].

The table shows the times required to print these models. The changing and key parameters in this optimization process were the layer height and the number of outer perimeters. As we can see the table interprets three basic series of production time conversion. These relate to layer heights of 0.264, 0.564, and 0.764 mm. Subsequently, two simulations are performed for each of these modifications. The first is with a filling percentage of the model of 20%, and the second with the absence of a pattern infill, which consists only of the already mentioned perimeters with a 20% overlap.

Variant No. 5 proved to be the most durable variant; its flexibility and resistance to damage were demonstrable during the actual handling and assembly. In terms of

Table 1 Printing parameters and total print time

	Number of top/bottom layers	Number of outline perime-ters	Infill value [%]	Default print speed [mm/min]	Layer height [mm]	Final print time [min]
Attempt 1 (20%)	3	2	20	3000	0.264	172
Attempt 2 (100%)	3	99[a]	–	3000	0.264	208
Attempt 3 (20%)	3	2	20	3000	0.564	109
Attempt 4 (100%)	3	99[a]	–	3000	0.564	89
Attempt 5 (20%)	3	2	20	3000	0.764	75
Attempt 6 (100%)	3	99[a]	–	3000	0.764	63

[a] An excessive number of outline perimeters was deliberately chosen. The program then recalculates and uses them to the extent that it ensures 100% filling of the model

time efficiency, the ideal was attempt no. 6; using these settings, it was possible to print the entire set of three models in 1 h 3 min.

3 Results

It is well known that the design quality of commercially available FDM devices is often debatable. Thus, device-specific settings often consist of unique modifications that may not apply to every device. However, this chapter presents a software solution and error analysis during the optimization of these processes, which are characteristic in specific cases of efforts to rapidly produce thin-walled components (Fig. 10).

If we want to evaluate the overall shortcomings that we encountered during the printing of shape shields, it is necessary to choose the right method. The most suitable method is the cause-and-effect diagram, also called the Ishikawa diagram. Diagram C-E was processed for FDM conditions of shape shield printing.

Figure 11 shows us the individual errors we encountered during component printing. As for the printing of plastic nut covers for screws, their production was seamless. Their composition is solid so the chosen material is ideal for their production. As can be seen from Fig. 11, most of the shortcomings occurred mainly on the frame of the face shield. The reason is, as already mentioned, its thin-walled design.

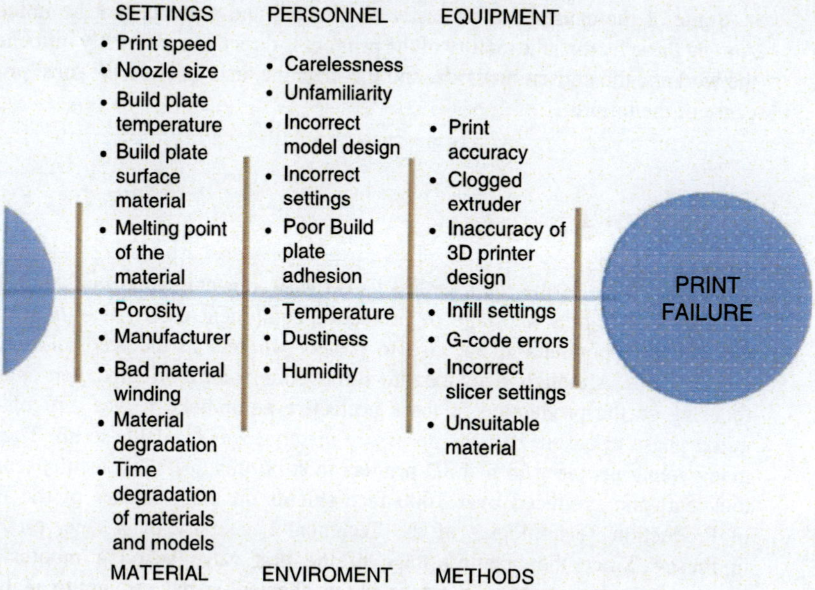

Fig. 10 Diagram C-E

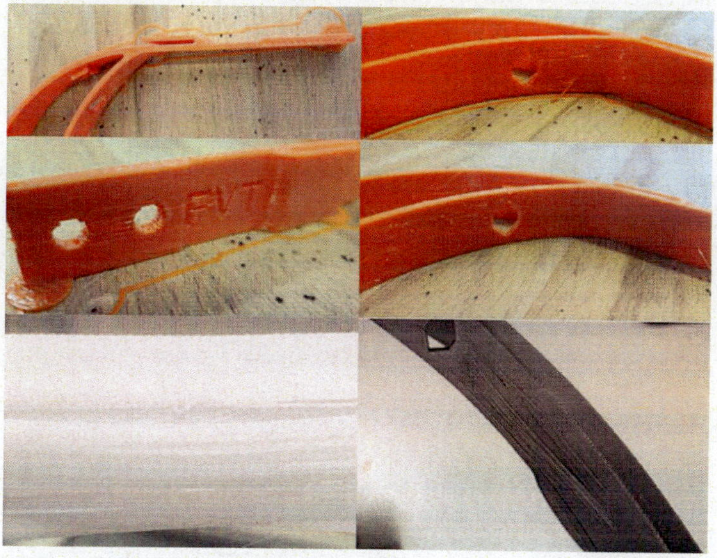

Fig. 11 Examples of defects in 3D printing

Some of the causes can be easily eliminated, and some cannot be influenced. Among these factors, the quality of the printout is directly or indirectly influenced by the workers, the chosen methods, and the machine, and they are the most probable cause of the failure.

4 Conclusion

In the first wave of the fight against COVID-19, our company fought for several months with a great shortage of protective equipment used not only to protect the country's population, but also to protect workers in the medical, social, or other spheres. Although many manufacturing companies reoriented their operations focusing on the production of these protective equipment, it was difficult in the initial phase to ensure that the necessary material was available to all. Therefore, many young people who had 3D printers took on this task. The faculty team also took part and produced over 1000 face shields on the premises of the Faculty of Production Technologies of the Technical University in Kosice, established in Prešov. Since the opening stage of the first wave lasted 4 months, many improvements were made. Therefore, in this chapter, we have sought to analyze the shortcomings identified during this period using a cause-and-effect diagram. Also on the basis of the analyzed process of 3D printing and assembly, it was possible to shorten from the original 4 h (as seen in Table 1) to the final time something around 1 h. As a result, the overall model before and after optimization changed minimally, but all user-centric requirements were maintained without changing the technical characteristics. As a main advantage of optimization, we consider that the total time has been reduced four times, all failures have been eliminated, and this has allowed us to produce larger numbers of face shields and help more people in the first line.

Acknowledgements This work was supported by the Slovak Research and Development Agency under the contract No. APVV-19-0590, by the projects VEGA 1/0700/20 and KEGA 055TUKE-4/2020 granted by the Ministry of Education, Science, Research and Sport of the Slovak Republic and by the Project of the Structural Funds of the EU, ITMS code: 26220220103.

References

1. J. Kaščak et al., Polylactic acid and its cellulose based composite as a significant tool for the production of optimized models modified for additive manufacturing. Sustainability **13**(3), 1256 (2021)
2. D. Perakovic et al., Internet of Things Concept for Informing Visually Impaired Persons in Smart Factory Environments, in *Industry 4.0: Trends in Management of Intelligent Manufacturing Systems*, (2019), pp. 69–86
3. V. Sabolova et al., Leaning of processes and improving the working conditions of the newly created working zone. Acta Logistica **7**(4), 283–290 (2021)
4. L. Roselli, *Green RFID Systems* (Cambridge University Press, Cambridge, 2014)

5. L. Gregurič, History of 3D Printing: When was 3D Printing Invented? https://all3dp.com/2/history-of-3d-printing-when-was-3d-printing-invented/. Accessed 21 May 2021
6. M. Brown, History of 3D printing. https://www.cadcrowd.com/blog/a-brief-history-of-3d-printing/. Accessed 25 June 2021
7. The Free Beginner's Guide. https://3dprintingindustry.com/3d-printing-basics-free-begin-ners-guide#03-technology. Accessed 18 June 2021
8. J. Grabara et al., Impact of legal standards on logistics management in the context of sustainable development. Acta Logistica **7**(1), 31–37 (2020)
9. M. Kliment et al., Production efficiency evaluation and products' quality improvement using simulation. Int. J. Simulat. Model. **19**(3), 470–481 (2020)
10. What is the influence of color, printing speed, extrusion temperature and ageing on my 3D prints? https://my3dmatter.com/what-is-the-influence-of-color-printing-speed-extrusion-temperature-and-ageing-on-my-3d-prints/. Accessed 27 June 2021
11. 3D printing infill: The Basics – Simply Explained. https://all3dp.com/2/infill-3d-printing-what-it-means-and-how-to-use-it/. Accessed 27 June 2021
12. A. Behunova, M. Behun, L. Knapcikova, Simulation software support of manufacturing processes in engineering industry. TEM J. **7**(4), 849–856 (2018)
13. K. Židek et al., Digital twin of experimental smart manufacturing assembly system for industry 4.0 concept. Sustainability **13**(9), 3658 (2020)
14. J. Kaščak, Research and implementation of an innovative approach to the design solution of device for additive FDM Technology, Technical University of Kosice, Dissertation work (2020), p. 95
15. L. Knapčikova et al., Material recycling of some automobile plastics waste. Przem. Chem. **95**(9), 1716–1720 (2016)

The Utilization of Progressive Methods in the Manufacture and Maintenance of Aircraft Components

Patrik Šváb, Peter Korba, Samer Al-Rabeei, and Jaroslav Hura

1 Introduction

The aerospace industry uses a wide range of materials, from metallic materials to plastics, ceramics, and composites. When building aircraft components, the optimal properties of the material must be carefully chosen. Each material must meet not only functional but also price and technological requirements. The functional properties of materials are used in aircraft constructions as much as possible. The materials are chosen mainly with the lowest possible specific weight, but at the same time with high specific strength. The strength characteristics of the materials are usually chosen regarding safety up to the maximum possible limit [1]. However, this means that the strength values of said material must be sufficiently reliable. The fatigue strength properties must also not be significantly reduced during long-term operation. This is associated with the overall life of the aircraft [2].

In terms of cost requirements, the price of construction materials must not be too high, because the higher price of composites is their major disadvantage, which prevents their greater development, not only in aviation but also in other industries that are engaged in the production of vehicles. On the other hand, as we can see on the Boeing 787 aircraft, where composites replaced aluminum on the fuselage and wings and their share in weight is up to 50% of the total weight of the aircraft, it is thanks to them that the aircraft is up to 1/5 lighter. This fact, together with the deployment of new engines and aerodynamic improvements, makes it possible to reduce fuel consumption by up to 1/5, which is currently quite significant at high fuel prices and low ticket prices. In the aerospace industry, building materials are defined as materials that are used in the manufacture of aircraft components.

P. Šváb (✉) · P. Korba · S. Al-Rabeei · J. Hura
Faculty of Aeronautics, Technical University of Košice, Košice, Slovakia
e-mail: patrik.svab@tuke.sk; peter.korba@tuke.sk; samer.abdo@tuke.sk; jaroslav.hura@tuke.sk

L. Knapčíková, D. Peraković (eds.), *6th EAI International Conference on Management of Manufacturing Systems*, EAI/Springer Innovations in Communication and Computing, https://doi.org/10.1007/978-3-030-96314-9_11

They are characterized by chemical composition and various properties, which are always destined for a specific use that is specific to the material. Aircraft technology is used in operation in conditions that are very diverse [3]. Therefore, the construction material must be selected according to the required properties based on the complete formulations of the file. The main interest in the construction and maintenance of aircraft is the general properties of metals and their alloys such as hardness, malleability, ductility, flexibility, strength, density, brittleness, meltability, and conductivity contraction and expansion.

Hardness refers to the ability of a material to withstand wear, penetration, cutting, or permanent deformation. Hardness can be increased by cold working metals, in the case of steel and some aluminum alloys by heat treatment. The components are often made of metals in their soft state, and then heat-treated to strengthen them, thus preserving their final shape. Hardness and strength are closely related properties of metals. One of the most important properties of materials is strength. Strength is the ability of a material to resist deformation. Strength is also the property of a material to withstand impact without damage. The types of loads or shocks to which the material is exposed affect its strength.

Density is the weight per unit volume of material. It is also called specific weight. When working with aircraft, the specified weight of material per cubic inch is preferred because this information can be used to determine the weight of the part prior to actual production. Density is an important factor in selecting the material to use in the design of a part to maintain the correct weight and balance of the aircraft. Metal which can be forged, rolled, or pressed into various shapes without cracking, breaking, or leaving any other harmful phenomenon is considered to be malleable. This property is required for sheet metal that is processed into curved shapes, such as covers, fairings, or wings [4]. Copper is an example of wrought metal. Ductility is a property of a metal that allows it to be permanently stretched, bent, or twisted into various shapes without breaking. This property is important for metals for use in the manufacture of wires and pipes. Formable metals are highly preferred in aircraft manufacturing because they are easy to form and resistant to impact loading errors. For this reason, aluminum alloys are used for the engine cover, fuselage, wing surface, and shaped or extruded parts such as ribs, beams, and bulkheads in aircraft. Chrome molybdenum steel is also easily molded into the desired shapes. Ductility is similar to malleability.

Flexibility is a property that allows the metal to return to its original size and shape when the force that causes the shape to change is removed. This feature is extremely valuable because it would be highly undesirable to have a component that is permanently deformed after the load is removed. Each metal has a point known as the limit of elasticity, and it cannot be loaded above this value, as it would cause its permanent failure. In the construction of an aircraft, the components and parts are designed so that their maximum load to which they are subjected does not exceed values above their limits of elasticity. This desired property is present in spring steel. A material that is tough, able to withstand tearing or shearing, can also be stretched or otherwise deformed without damage [5]. Toughness is a desirable property for the metals contained in the aircraft structure. Brittleness is a property of metal that allows only small bending or deformation without damage. Fragile metal is able to break or crack without changing shape. Because structural metals are often subjected to impact loads, brittleness is not a highly desirable property. Examples of brittle metals are cast iron, cast aluminum, and very hard steel.

Meltability is the ability of a metal to become liquid when heat is applied. During welding, metals are melted. Steels melt around a temperature of 1426.667 °C and an aluminum alloy at about 593.33 °C. Conductivity is a property that allows a metal to transfer heat or electricity. The thermal conductivity of a metal is especially important in welding because it regulates the amount of heat that will be needed for proper melting. The conductivity of the metal to some extent determines the type of preparation to be used to control expansion and contraction. In an aircraft, electrical conductivity must also be considered in conjunction with coupling to eliminate radio interference. Thermal expansion refers to contractions and elongations that are reactions produced in metals as a result of heating or cooling. The heat applied to the metal causes it to enlarge. Cooling and heating affect the design of welding jigs, castings, and tolerances required for hot-rolled material.

2 Materials Used in Aviation

Steel is an alloy that is produced by combining iron and carbon, the content of which is up to 2%. Carbon has decisive effects on the properties of ferrous alloys. In addition, other elements are applied to a significant extent. Some of the presence in steels always occurs in connection with the raw materials that are used in the process and with the production process itself. We can name them accompanying elements. Additional elements are added to the steels to adjust their properties in the direction required. We call this element additive or alloy (respectively alloying) elements. Usually, the accompanying elements are divided into harmful or beneficial. Components and parts of complex shapes are produced by casting into molds. This method is often the most cost effective, especially for machine parts that are less stressed [6]. The bodies of machines are made in the form of casting. These are, for example, turbines, engine parts of vehicles, or electric motors and others. Cast iron is the most used material in the production of castings. It is an alloy of iron and carbon with a content of more than 2.06%. Gray and white cast irons are the main representatives of cast irons. Other types that are used are derived from these types of base cast irons. Steel is another material used in the manufacture of castings in addition to these materials.

Except for iron, all other metals are nonferrous but, like iron, are almost unusable in pure form for structural purposes. We meet them practically only in the form of alloys. Copper is one of the exceptions as it is used for electrical purposes. Nonferrous metals are subdivided according to their specific gravity into:

- Lightweight metals have a density of less than 5000 kg m^{-3} and are most important for the aerospace industry. They are mainly alloys of aluminum, magnesium, and titanium. They have the highest share in aircraft with a classic metal structure, especially in thin-walled elements of structures.

- Heavy metals have a density higher than 5000 kg m^{-3} and are used in the aerospace industry only in some elements of the structure, namely in airframe systems that are outside the primary structure. These are mainly alloys of copper, zinc, tin, lead, nickel, cobalt, and antimony.

Copper is a red metal that crystallizes in a surface-centered cubic system. It has a specific gravity of 8.93 kg/dm^3 and its melting point is 1083 °C. The tensile strength at 40% elongation is 220 MPa (rolled copper). With the help of cold forming, it is possible to increase the strength up to 400 MPa. Impurities have a very adverse effect on the mechanical properties, in particular oxygen, which causes embrittlement at a content of more than 0.1%. The ductility of copper is good both hot and cold. It is also a material that welds well and solders well [7]. It has very good resistance to weathering, as it is covered in air by a layer of compounds (carbonates, sulfides, oxides), which protect it from further possible corrosion and thus cause discoloration of the surface. It is also very resistant to seawater and some acids except nitric acid. Sulfur and ammonia have adverse effects on copper, which severely degrades it.

Aluminum alloys are an important construction material that have particularly better mechanical properties than pure aluminum. They are therefore a construction material that is very important and is increasingly replacing other classic materials in constructions, including steel. Significantly, this dominant predominance of aluminum and also its alloys can be seen in aircraft production. Aluminum alloys can be divided into two groups according to the method of its processing, namely alloys intended for forming or casting [8]. Both groups have in common the fact that the content of their additive elements is the same and that some of the alloys located in the range of both groups can be used as alloys intended for forming or casting.

Nickel is a metal of medium hardness and can be polished well. It has a specific gravity of 8.9 kg/dm^3 and crystallizes in a cubic lattice that is area centered. Its melting point is 1453 °C and it has ferromagnetic properties up to 356 °C. The strength of nickel varies from 300 to 500 MPa, depending on the form of its processing. It has good formability in hot and cold conditions. Nickel strength can be increased by cold forming up to 1100 MPa. The amount of nickel (about 60%) is currently used in alloy steels. About 25% is used to produce nickel alloys and 15% is used in the production of pure nickel semifinished products. A certain amount of nickel in the pure state represents anodes which are intended for surface treatment by nickel plating. It has good resistance to atmospheric corrosion and water (including marine). It also has very good stability at higher temperatures up to about 800 °C.

Special metallic materials called precious metals are gold and silver, which are most often used only as parts of some electronic/avionic elements of aircraft. It can be expected that in the future, sintered metallic materials, which result from the heat treatment of pressed pure metal powders, will be used more. Elements that are complex in terms of strength can thus be produced directly and in a targeted manner also influence the internal structure of the materials [9]. At the beginning of aviation, the most important were natural materials, the largest representative of which was wood. This material was available and easy to process, so it spread relatively quickly even in systems that are large in space and strength. In the form of wooden parts, which are mechanically machined, we now rarely encounter wood, practically only in aircraft of very light categories. Rubber is also still a material that is traditional to produce tires, seals, hoses, and spring elements.

2.1 Composite Materials

Composite materials are modern materials, the development of which began not so long ago, even though their construction itself was observed, to be honest, from nature. Among composite materials with a fibrous structure, we also include wood, for its structure, but also bamboo, grass blades, or structure of animal bones. Composite materials are divided according to their type of reinforcing component into composites that are reinforced with fibers or short fibers (so-called whiskers), composites that are reinforced with dispersion (e.g., metal oxides), and composites produced by controlled crystallization, i.e., eutectic composites. Composites are further divided according to the type of matrix into composites with organic matrix (polymer), ceramic, and metal. The last group consists of composites formed by a silicate matrix (prestressed concrete or reinforced concrete), which are often used in construction [10].

The composite is most often taken to be a material that meets the conditions that it was artificially created, its composition consists of at least two components that are significantly chemically different from each other, its components have an even distribution throughout its volume, and its final properties are different from the properties of the used components. Depending on these resulting properties, composites are divided into:

- Composites that have high mechanical properties
- Composites that have special physicochemical properties

ARALL was the first fiber-metal laminate used in aerospace. It was invented in 1978 in laboratories designed for materials research in the Netherlands. This laminate, made of sheets of aluminum alloys of type 2024 and 7475, respectively, and epoxy prepregs of aramid fibers, was designed in a fiber arrangement in one direction with an orientation towards the dominant load. The first two patents for this material were recognized internationally in 1984, and after gaining confidence, up to four types of this type of laminate were subsequently produced. After considering the use of the newly formed composite material, it was found after evaluations in flight simulations that by using it, the weight of aircraft can be reduced by up to 30% and their service life can be significantly extended. Due to its excellent fatigue resistance, ARALL began to be used mainly on the lower parts of the wings for covers and rear rudders. However, a thorough study has shown limitations that have been undesirable. In particular, due to the low adhesive strength between the resin

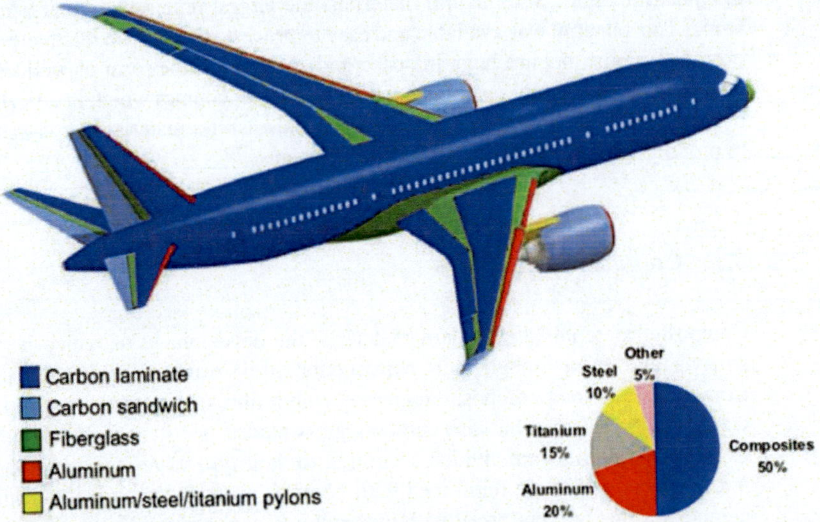

Fig. 1 Representation of materials in the Boeing 787-8 aircraft

and the fibers, the maximum value of the number of fibers (about 50%) was very low in the matrix. Another problem was that ARALL cannot be made with fibers that are oriented in multiple directions. These problems were fundamental as they limited the laminate properties and thus its use in places where it is subjected to high stresses (for example in the fuselage). Therefore, the research had to continue. Representation of materials used in the Boeing 787-8 aircraft can be seen in Fig. 1.

Instead of aramid fibers, glass fibers were first used and applied in the laminate in 1990. This laminate was called GLARE and solved many of the negative aspects of ARALL. In addition, it surpassed its many mechanical properties. GLARE could finally be processed with the orientation of the fibers in multiple directions. This allowed the designers to use this laminate in a much larger proportion of structures. Currently, GLARE is used by one of the largest aircraft manufacturers, Airbus, which uses this laminate extensively in the construction of the A380 transport aircraft for the upper part of the pressurized cabin. It is also used on the lower covers of the wings, rudders, and many other places. However, these refined materials are not the end of the development of fiber-metal laminates. In the future, these new composite materials have a potential use in the production of an increasing number of aircraft parts, regardless of the fact that they can be used in other areas of industry (space, ship, automotive, sports, or construction).

3 Requirements on Materials in the Aerospace Industry

There are certain basic requirements for materials used in aviation. These requirements, due to their high energy consumption, are high strength at low weight and resistance to stress. Resistance to atmospheric conditions is also very important. These conditions include rain, hail, icing, atmospheric discharges, heat stress, sunlight, or birds. In particular, the strength requirements are formulated by aeronautical regulations to minimize the likelihood of failures due to the lower value of the strength characteristics of the material than those taken into account in the design calculations.

3.1 Load Resistance

If either dynamic or static tests are used to demonstrate compliance with the airworthiness regulations, factors suitable for repair shall be used to correct the test results depending on the construction materials used. Structures that have a safe service life (statically determined) have a number greater than 1.0. In the case of fail-safe structures, i.e., statically determined, the test results do not need to be corrected.

3.2 Selection of Suitable Material

The suitability as well as the durability of the materials used for the production of components, the failure of which may adversely affect safety, is assessed on the basis of demonstrable tests and experience. This corresponds to the approved standards, which are consistently applied. It is the technical regulations or military and civil aviation specifications that ensure that the actual load-bearing characteristics of the structures as well as other properties will be in accordance with the design data. In addition, all probable environmental influences, such as humidity or temperature, that can be expected when operating aircraft must be considered. The strength characteristics of the material must be documented by the required quantity and by the required quality of the performed tests, processed by means of statistical methods.

The calculations must use values of the type whose probability minimizes structural failures due to scattering of material properties. For design calculations, it is possible to apply only such values of material characteristics that guarantee that:

- The values used provide strength characteristics with a probability of up to 99% at the specified level of 95% in the case of a structure which has a concentrated load applied to one strength element in which its failure would cause a loss of system integrity

- The values used provide strength characteristics with a probability of up to 90% at a given level of 95% in the case of a structure which has a load distributed through a larger number of strength elements of the system
- Where temperature effects may be important under normal operating conditions, the temperature effects on the strength characteristics of the material used in the calculation of the primary structure must also be considered
- The possibility of catastrophic fatigue failure must be minimized by strength, detailed calculations, and design, especially in places where concentration of the structure is possible

The use of strength characteristics with higher values is possible for calculation purposes, if a sample of the selected material is always tested before each use of the material for the production of the strength element so that the assumption of real strength characteristics is equal to or even higher than that used in the calculation. For the calculations, it is in principle possible to use such values of the characteristics of the materials which have been obtained based on compliance with the standards which guarantee the stated material mechanical properties by performing tests on samples of each material which has been applied. Only in this case is it possible to use a coefficient of 1.0 for strength calculations. Material strength characteristics and selection depend on which airplane part will be made of it. Summary of airplane parts and their functions is shown in Fig. 2. If the material characteristics used are only in their values from the tables, a correction factor of 1.25–1.5 must be used for the strength calculations.

When put into service, large-capacity aircraft also brought with them very special requirements designed to create the safety of crews and passengers themselves in the event of an accident with the subsequent occurrence and spread of fire. Construction materials for the construction of aircraft structures and fuel tanks should not produce fumes that are toxic or suffocating during combustion. They should also not occur when the aircraft is extinguished with aircraft extinguishers. The fire should spread so slowly that all passengers, as well as the crew on board the aircraft, have a chance to survive a critical situation and escape from the aircraft in an emergency within the specified limit.

In all places in the aircraft structure where flammable liquids may be present and where flammable liquids may escape from the aircraft systems, the possibility of ignition of these liquids or vapors and the possibility of hazardous consequences in the event of fire must be minimized. Demonstration of this must be in the form of tests or analyses:

- The potential sources and locations of flammable substances and how such potential leaks occur are detected
- Characterization of combustibles, which also includes the burning effects of liquids or the materials themselves that absorb these liquids
- Sources of ignition that are possible, including electrical failures, overheating of equipment, and possible failure of prevention equipment

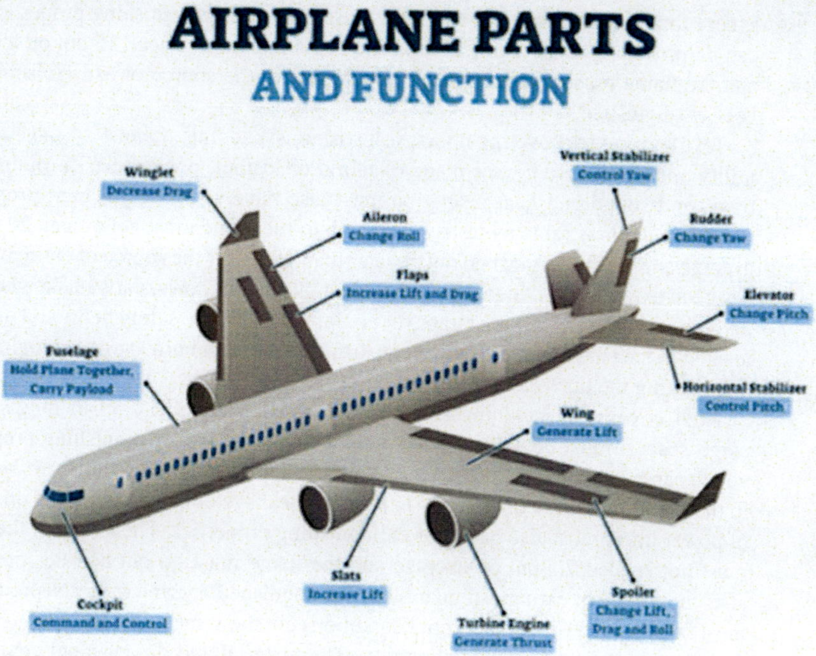

Fig. 2 Summary of airplane parts along with their functions

- The availability of means to suppress and subsequently extinguish the fire, such as interrupting the flow of flammable substances, switching off the necessary equipment, and use of fire barriers or a fire extinguisher
- Dispositions of aircraft components such as fire resistance and temperature effects

Flammable liquids and vapors must be defined and identified in detail from each area of their possible escape. Parts of the engine installation, basic control system, and also parts of the primary structure which occur in areas where leakage of liquids is possible, which are flammable or in areas which are adjacent and could thus be exposed to direct fire, must be manufactured from fire-resistant construction material or must be shaded in such a way as to be able to withstand fire and its effects.

3.3 Material Test Criteria

Construction materials that are used to produce ceiling panels, galleries, partitions, floors, and large cabinet walls are also used in the construction of storage rooms, except for subfloor compartments, and must have self-extinguishing properties that can be proven by tests. The length of the flame must not exceed 15 cm on average and the flame must go out no later than 15 s after the source of extinguishing has been extinguished.

Textiles used for covering floors, seat cushions, padding, upholstery and curtains, utility and decorative covers made of fabric or leather, power lines, furniture and trays, and sound and heat insulation must also have self-extinguishing properties which are demonstrable by tests. The length of the flame must not exceed 20 cm on average and the flame must go out no later than 15 s after the source of extinguishing has been extinguished. Signaling panels and plastic windows, parts made wholly or partly of elastomers, lashing straps for equipment or cargo, safety belts, and luggage located in the passenger and crew compartments may not burn at a speed higher than 6 cm during the horizontal test per minute.

Roof coverings, including sound and thermal insulation used in luggage and cargo spaces, must be made of materials which have self-extinguishing properties which can be demonstrated by tests. The same applies to the formation of flames as in the previous case. Components in the electrical system, i.e., insulation on cables or power lines, must also have self-extinguishing properties. The length of the flame must not exceed 7.5 cm on average and the flame must go out no later than 30 s after the source of extinguishing has been extinguished. During the tests, procedures must be provided so that the initial conditions are the same as for the fuselage of an aircraft in an air-conditioned condition. The material and dimensional conformity of the samples must be the same as the actual ones used in the aircraft structures. For the test, they are placed in a metal frame and exposed to the flames of a burner at a temperature of 845 °C in the center of the flame.

4 The Potential of Progressive Methods in Aerospace Industry

Additive manufacturing (AM) can be defined as the exact opposite of traditional machines to produce parts or components. Several AM techniques have been introduced in recent years. One way to classify these techniques is based on the state of the raw material used, which may be liquid (FDM) or discrete particles, usually powders, or solid sheets may be used. FDM is a relatively inexpensive technique based on applying a thin plastic wire along a predetermined path: the wires are melted in a nozzle that describes the movement according to a digital replica of a model in the 3D plane relative to the shape of the workstation. After leaving the nozzle, the plastic solidifies, and it is possible to obtain a solid model. The SLA technique is based on a photosensitive liquid resin that solidifies after being struck by a laser beam. Even in this case, layer by layer, a solid shape can be obtained by polymerizing the liquid along the nozzle path. This can be useful for obtaining nonstructural parts. What may seem extremely complicated at first glance is, in a simplified form, the creation of parts and components on 3D printers that are specially adapted to the needs of a particular industry. However, if high-strength materials are required, methods based on melting metal powders should be used. Solid parts with structural properties comparable to the workpiece or cast metal are obtained from aluminum, steel, or titanium powders. Finally, laminate object modeling (LOM) is based on the application of thin sheets of paper, metal plastic, or adhesive-coated laminates until a 3D shape is obtained that connects the individual layers.

What makes AM interesting is that, thanks to the maximum freedom in shaping, it is possible to create even extremely complex shapes. Complex structures based on thin lattice structures can achieve high structural efficiency, where the whole material exhibits similar stress values at all points. The Industry 4.0 concept introduces several key assistive technologies that could change aircraft maintenance. Networking, availability of big data, ability of delocalized and personalized production, networks of interconnected microsensors, intelligent and intuitive visualization of information in remote operations, and automation are examples of technologies suitable for use not only in factories but also in aviation.

However, efficient hardware processing requires efficient big data processing algorithms and the support of local real-time monitoring strategies for composite structures, which manufacturers such as Airbus and Boeing already use to improve aerospace monitoring. The aim of this approach is to collect data from millions of sensors to monitor the structure and detect microcracks as efficiently as possible, enabling a so-called damage tolerance strategy. Technologies such as AM and AR, which have been improved by the Industry 4.0 revolution, will be a key element in the aviation maintenance strategies of the future. It should be noted that aircraft, helicopters, and other automatic machines can be composed of millions of parts that require complicated maintenance, which often require specific equipment and tools needed for assembly or disassembly. VR can be useful in these areas of advanced engineering, where it is possible to undergo virtual training and practice precise procedures in various situations that reduce the occurrence of errors due to incorrect placement and type of parts. A realistic 3D CAD model of the part that needs to be maintained can be projected into a video stream, where an external organization, such as a technician specializing in these types of failures, can also comment on the problem. Training in virtual reality, where the scenario moves forward only if the set conditions are met, can help reduce the risk of skipping a phase or step of maintenance, which can very easily occur in a real situation.

Additive production can be useful if large stocks need to be avoided. The part can be made of metals such as aluminum or titanium, provided that a suitable machine and powders are available. In addition, if redesign of the part is possible, weight reduction can also be achieved with optimized grid structures. AR can help operators with user-friendly and transparent manuals where virtual models and instructions are mixed with the real world. In such a case, a reduction in the workload and time required to complete the tasks and an increase in reliability can be expected as a consequence of the reduction in the number of errors caused by opaque and in some cases often available manuals. AM can also be combined with AR and

VR. In fact, VR can be useful for creating an extended maintenance manual and illustrated parts catalogs, where the appearance of the part being repaired is designed by the operator intuitively based on the actual aircraft. This is especially true for VR- and AR-based manuals, where assembly and disassembly tasks are presented to operators through a combination of CAD models, symbols that suggest manual operations, and virtual panels where operations can be controlled and communicate with technology through gesture tracking. AM could also be used for remote maintenance, where virtual animations could be prepared in real time to support and manage complex procedures through centralized maintenance centers and, if necessary, the captured image would be projected to remote operator access devices. This could overcome one of the problems that currently limit the use of VR in industry: the time needed to prepare animations. In the case of a modern transport aircraft made of millions of parts, it should be noted that the implementation of animated virtual sequences for each possible maintenance operation would be time consuming. Training is another task where augmented reality could increase realism and efficiency.

5 Conclusion

The ability of AR and VR to combine virtual and real parts allows you to simulate complex scenarios without the need for inaccessible parts. Overall, the VR's impact on aircraft maintenance can be huge, especially in dealing with complex operations that cannot be clearly explained in traditional paper manuals. In addition, the documentation must be in operation for several decades, with the average service life of a commercial aircraft ranging from 20 to 30 years. The VR can help to update maintenance procedures for instructions based on textual, real-time maintenance documents. However, implementing all commercial aircraft maintenance tasks in the VR can be a complex task [11]. At present, the most realistic solution is to design the initial implementation of only the most critical and demanding maintenance operations in the VR. The integration of short-term maintenance preparation environments and enriched scenes into the VR system using CAD modeling could reduce the time required to implement the entire maintenance manual in a virtual environment in the near future [12].

VR can be used in aviation to support maintenance in several ways. Illustrated AR- or VR-based parts catalogs, virtual environment-based maintenance manuals, VR-based remote maintenance software tools, and VR-supported assembly and disassembly operations are just one of the many solutions this technology provides that can benefit from implementation of virtual reality to the aviation industry. This technology can be useful for bridging the gap between traditional text-manual training, where simplified two-dimensional sketches represent complex components and reality, where complex 3D shapes must be detected and assembled, often hidden under a fairing or other components and assembled using advanced technology tools.

References

1. Y.I. Buryak, M.P. Lyubovnikov, Traceability of aircraft components state in real time. Vestnik Komp'iuternykh i Informatsionnykh Tekhnologii **191**, 32–42 (2020). https://doi.org/10.14489/vkit.2020.05.pp.032-042
2. A. Hobbs, Aircraft Maintenance and Inspection, in *International Encyclopedia of Transportation*, (2021), pp. 25–33. https://doi.org/10.1016/b978-0-08-102671-7.10103-4
3. R. Janeliukstis, D. Mironovs, Smart composite structures with embedded sensors for load and damage monitoring—a review. Mech. Compos. Mater. (2021). https://doi.org/10.1007/s11029-021-09941-6
4. L. Jiangshan, C. Ming, Research and application of virtual simulation technology in the aerospace bearing design and manufacture. MATEC Web Conf. **151**, 04002 (2018). https://doi.org/10.1051/matecconf/201815104002
5. E.N. Kablov, V.V. Antipov, J.S. Oglodkova, M.S. Oglodkov, Development and application prospects of Aluminum–Lithium alloys in aircraft and space technology. Metallurgist (2021). https://doi.org/10.1007/s11015-021-01134-9
6. C.S. Karunakaran et al., Reliability augmentation through technological applications in Indian aircraft maintenance training sector. Turkish J. Comput. Math. Educ. **12**(11), 1964–1970 (2021). https://doi.org/10.17762/turcomat.v12i11.6151
7. E. Lumnitzer, M. Andrejiová, A. Goga Bodnárová, Verification of the impact of the used type of excitation noise in determining the acoustic properties of separating constructions. Measurement **78**, 83–89 (2016). https://doi.org/10.1016/j.measurement.2015.09.030
8. S. Mohd Yusuf, S. Cutler, N. Gao, Review: the impact of metal additive manufacturing on the aerospace industry. Metals **9**(12), 1286 (2019). https://doi.org/10.3390/met9121286
9. H.Y. Pai, in *Mixed Reality of Augmented Reality in Mobile Learning for Aircraft Maintenance*. 2020 IEEE Eurasia Conference on IOT, Communication and Engineering (ECICE) (2020). https://doi.org/10.1109/ecice50847.2020.9301992
10. S. Pantelakis, K. Tserpes (eds.), *Revolutionizing Aircraft Materials and Processes*, 2nd edn. (Springer, Cham, 2020). https://doi.org/10.1007/978-3-030-35346-9
11. M. Piňosová, M. Andrejiova, E. Lumnitzer, Synergistic effect of risk factors and work environmental quality. Qual. Access Success **19**(165), 154–159 (2018). https://www.researchgate.net/publication/326849276_Synergistic_effect_of_risk_factors_and_work_environmental_quality
12. E. Sentyurin, I. Mekalina, M. Aizatulina, Y. Isaenkova, The history of aircraft materials of glass and polymer materials with special properties (To the 75th anniversary Laboratory of polymer materials with special properties). Aviat. Mater. Technol. **3**, 81–86 (2017). https://doi.org/10.18577/2071-9140-2017-0-3-81-86

Part IV
Education and Awareness of Sustainability

The Specifics of the Silver Generation Shopping Behavior in the Selected Cities of Western Slovakia

Alena Dubcová, Miroslava Trembošová, L'udmila Nagyová, and Dagmar Cagáňová

1 Introduction

In addition to globalization and internationalization, the dynamic development of the knowledge-based economy and the stimulation of innovation, population ageing is also considered to be a major socio-economic trend in society [1–4]. Ageing is not only a biological process but also a multidimensional and diverse phenomenon that depends not only on human beings but also on the determinants of their biological, economic, social, personal, psychological, cultural and historical environment. The issue of an ageing population has been of interest to the EU for the last two decades. It is dealt with in the Ageing Report, which is published every 3 years by the Working Group on Ageing of the Economic Policy Committee (EPC) and the Directorate for Economic and Financial Affairs (DG ECFIN) of the European Commission under a mandate. The latest [5] was carried out in 2017. Ageing issues are covered in various discussions and processes at the European Union level, including the overall strategy of [6] to ensure smart, sustainable and inclusive growth. It reflects the quality of the individual EU countries' citizens' life. Since most European countries are after the second demographic transition, 2012 has been

A. Dubcová · M. Trembošová
Constantine the Philosopher University in Nitra, Nitra, Slovak Republic
e-mail: mtrembosova@ukf.sk

L'. Nagyová
Slovak University of Agriculture in Nitra, Nitra, Slovak Republic
e-mail: ludmila.nagyova@uniag.sk

D. Cagáňová (✉)
Faculty of Materials Science and Technology in Trnava, Slovak University of Technology in Bratislava, Trnava, Slovak Republic
e-mail: dagmar.caganova@stuba.sk

L. Knapčíková, D. Peraković (eds.), *6th EAI International Conference on Management of Manufacturing Systems*, EAI/Springer Innovations in Communication and Computing, https://doi.org/10.1007/978-3-030-96314-9_12

declared the European Year for Active Ageing to promote healthy ageing by the European Union. An older but healthy and active person is more beneficial to the society than his/her passive opposite.

Decrease in birth rate, emigration of young people and prolongation of human life are reflected in changes in the age structure of the population and in the Slovak population. The share of older age cohorts in the total population is increasing. According to the middle population scenario of the largest cities in Slovakia in Nitra and Trnava, more than 30% of the population over 60 years of age should live by 2035 [7] adding the younger segment of the older population (50–60 years), and the share of the silver generation in the population of these cities will reach more than 46%.

Older age is a term that does not have a specific age limit. Most often it is related to the population of post-production (50 and more) and post-productive age (65+); participants are also referred to as the silver generation. The silver generation (silver generation) represents a strong group of potential consumers; it is a basic silver market that develops across multiple exposures, showing no retail outlet where another manufacturer can be found, but rather as a customer. Today's older consumers are perceived as actors who are more wealthy, active and vulnerable to their health, which can find out who is all right and bother after death as the previous generation had [8].

The problems of ageing have attracted the general professional and scientific public in recent decades. The issue of population ageing is of interest to several scientific disciplines (economic, demographic, geographical, urban and other).

Ageing is undoubtedly not only a biological phenomenon, but also an inevitable, long, diverse and multidimensional process that depends on human beings on the one hand and on different determinants, economic, social, biological, psychological, cultural and historical, on the other side [9].

The aim of the chapter is to evaluate the selected aspects of silver-generation shopping behaviour—older customers aged 50 and over by selected domains related to shopping (e.g. location, price, atmosphere, product range)—and determine their types and specifics of shopping behaviour in two regional cities of Nitra and Trnava from the respondents' perspective.

2 Theoretical Background and Methodology

The population of the elderly is not homogeneous, but rather diverse not only in terms of age but also in terms of its characteristics and needs. The ageing population is perceived differently by the society. Some authors perceive the ageing population as a burden on society, characterizing it as a "silver tsunami" that is approaching and for which society is not prepared [10]. Generally, old age is perceived or depicted by society as a period of human characterized by inactivity and dependence. This public view of the older generation is starting to change as a result of its activity. Lifestyle is also an important factor in this change. The dynamic development and rationalization of lifestyle lead to the effort of the silver generation to cope with age—to "rejuvenate" or to imitate the behaviour of others, in many cases younger people (e.g. dressing, leisure activities) of the younger generation which is directly reflected in the shopping behaviour of this generation. This decision-making process for older consumers provides the basis for the development of the mega-trend known in the literature as a "rejuvenating population". The existing stereotype of older people as a lonely and poor person is increasingly being replaced by images of active people and those interested in an active life, whose approach to life is more hedonistic and who seek to meet their needs and the needs of their closest relatives. Therefore, older people are increasingly considered as an important market segment [11]. Some studies, e.g. [12], emphasize that the generation of baby boomers should be seen not only as consumers, employees or employers but also as a group of people that redefine the concept of older age. This group of people have different experience, values and expectations than their parents have had. They have changed at all stages of their lives, whether it is talked about family, work or education and are linked to the further active use of their free time in retirement.

Today's consumers are more informed and more aware of their needs and desires, and to a greater extent compare and consider all possible factors [13]. Older people no longer only consider economic aspects in their consumption but, on the contrary, price becomes only one of several factors [14], which is only possible with a stable and relatively higher income level of the older generation, which includes the incomes of seniors in the old EU countries and only a very low proportion of older people in the new EU member states, including Slovakia. In Slovakia, older people do not reach the pensions of seniors in Western Europe. Although pensions represent a guaranteed income, the Slovak seniors are often dependent on economically active children, who make a significant contribution to the care of their parents. In this view, the shopping power of many households currently does not allow to meet demand not only for food, but also for other basic products. Nagyová et al. [15] and Singh and Kathuria [16] point out that lower income people generally buy low-quality, unhealthy and non-branded products.

In the absence of a labour force, especially in the coming decades, older people will play an increasingly important role in employment, increasing their income. Casey [17] pointed out the possibility of "activating" older workers to increase economic growth. According to the World Health Organization [18], the process of creating opportunities and conditions for active maintenance of health and participation in life and society—maintaining social relations and achieving the highest possible level of independence and self-sufficiency of an individual—increases the quality of life of older people.

Older people also have an increasing share of consumption, increasing their impact on production and consumption patterns. According to Schaffnit-Chatterjee [19], the ageing of the population significantly affects the nature of demand within individual segments of goods and services.

Older people have to cope with several changes. The primary ones include biological; although they come gradually with age, they are evident in the older age. Ageing also contributes to psychological changes in different areas, changing cognitive abilities and functions, attitudes, moods and emotions. Change of cognitive abilities can lead to a change in memory or information-processing capabilities. There are changes in a person's mobility, flexibility or elasticity. All these changes have an impact on consumer behaviour and will be increasingly reflected on a market that adapts them goods and services as well as marketing and customer communication. Seniors have their typical habits regarding consumption, savings and investment, even if their income and expenses change with their social, professional and family roles. In Slovakia, the income and expenditure situation of seniors is characterized by low-income levels and high expenditure on food, consumer goods and health. Although the incomes of seniors in Slovakia have increased slightly in recent years as a result of gradual valorization of pensions, but in particular by increasing the number of seniors, silver households live with lower incomes and are unable to maintain the same standard of living as when they were professionally active [20].

At present, the silver generation consists mainly of the post-war generation known as the baby boomer—a generation born in 1946–1964, whose individuals reach the age of 55 or more. This generation is characterized by the survival of several transformations in the political, economic and social spheres, which is reflected in its consumption behaviour. It is often considered to be teenagers in a growing consumer society [21].

Unlike the previous older generation that was characterized by fragility and passivity is richer and used to consume and make decisions. Another important factor is its high level of education, which can lead to increasing demands for high-quality services. The generation of baby boomer is of interest to several authors. Kasčaková and Kubisová [20], who investigated changes in their consumer behaviour throughout their life cycle and their propensity to consume, found that this generation is not a homogeneous group, which is reflected in its consumer behaviour and inclination to consumption. Today's seniors are healthier, more informed and interested in being integrated into society more than ever [22]. Even though they are physically older, they feel mentally still young, to which they also adapt consumer behaviour. They present a "forever young" attitude to life. Leach et al. [22] shows that cosmopolitanism is the central part of their personality and also influences their consumer behaviour patterns, which suggests their strong tendency to travel (they are more experienced travellers than their parents); they are increasingly looking for sightseeing tours, their travel is associated with their lifelong hobbies, and they also realize various adventure tours. Most people in this generation think into the future, so their behaviour affects more how many years of their lives are left than they actually are (that is, they focus on things they do before they die). This also implies their attitude to property. Of course, this attitude varies from individual to individual, but more than other generations, the tendency for this generation to consume its property is higher than the tendency to pass it on to the next generations.

Another characteristic of this generation is that women live longer than men, on average up to 7–11 years. In Slovakia, life expectancy is 72.9 years for men and 80.3 years for women; women live longer than men by 7.4 years (2017). They are more vital and healthier. They increase social and family activities at retirement age, while men reduce them [23].

The baby boomer generation is often perceived as a “sandwich generation”, which has responsibilities not only to its children, of which up to 37% live in the same household with their parents, but also to their still-living parents, which also affects the redistribution of its wealth [21].

Other social changes include changes in the educational structure. There was a significant increase in the share of persons with full secondary and tertiary education and a decrease of persons with elementary education, but also higher women’s participation in the labour market and thus their better ranking in the labour market. Acquired higher incomes are also reflected in the consumer behaviour of older generations and thus also in their expenditure structure. Changes in consumer behaviour are also affected by other economic and social factors such as economic growth of the economy, size of net income of households, development of shopping power of households and development of savings.

Expenditure in the elderly population can also be reduced by changing the family structure, e.g. the independence of children, although the Slovak population is characterized by the late independence of children. In many cases, seniors live in one household with their adult children, sometimes unemployed, which reduces their income. In general, the average level of consumption of individuals as well as their standard of living decreases at retirement age, which is also reflected in the change in their consumption patterns.

The consumer behaviour of the silver generation is strongly influenced by the accumulated amount of savings for the creation of which individuals had different motives and conditions during economic activity. Many of them purposefully saved for “retirement” to finance their consumption at this age. The pension system in our society, with many older people, especially those who had low wages during their productive years, puts them all the way to poverty. People in the productive age are in debt until the age of 45; between the age of 45 and 65 they usually make savings, which they start to reuse after age 65. Many banks and financial institutions are already responding to this trend, e.g. by creating special accounts, insurance, etc. [21]. The older population constitutes an important and what is essential, growing group of consumers with increasing incomes. Its share in the total population is growing, the population has considerable shopping power of accumulated savings and increasing consumption is becoming significant and promising for the market.

According to Olejniczak [24] seniors are becoming one of the fastest growing segments of the consumer population, even though in the social and economic field there is a deeply rooted image of them as unattractive consumers for retail because of their limited shopping power and shopping, which are the result of social, psychological and health changes that occur with age. On the other hand, the shopping power of the silver generation is increasing and that is why the group of people over 50 is an important target group not only for retail, banking, tourism and culture, but also for other sectors. Olejniczak [24] thinks that the negative assessment of the position of seniors on the market is based on the lack of information about their internal characteristics and diversity. Research conducted by Emerald, ScienceDirect Elsevier, Jstor, Wiley Blackwell and others has shown an interesting picture of the behaviour of senior consumers:

- Seniors are interested in new products, but not for the satisfaction of specific personal needs (biophysical and psychosocial) in older age [25].
- Seniors with increasing age have limited ability to use additional information and rely more on memory experience [26].
- Older people are more sensitive to product prices [27].
- Seniors have a higher loyalty to retail outlets [28].
- Seniors prefer the convenience and simplicity of products [29].

Other authors such as Carstensen et al. [30], Williams and Drolet [31], Cagáňová et al. [32] and Vraňáková et al. [33] point to motivational changes in their decision-making. According to them, older people perceive the remaining time as limited and, from this perspective, prefer more social goals that are emotional and meaningful over those related to knowledge and with rationality. Often their routine consumer behaviour is habit-driven; they tend to favour long-established habits [34, 35]. According to them, older people are more loyal than younger consumers to their used brands. According to Cuddy et al. [36] the long-term stereotyped views of older people are quite difficult to change.

3 Methods and Data

Consumers are the driving force of the economy and an important factor in the exchange of goods, so knowledge of shopping behaviour is of great importance for the company and business especially in the retail sector. In order to keep pace with changing consumer demand, producers are bringing many new products to the market [37]. In a market economy, it is no longer a problem to produce a product, but it is much more difficult to sell a product. Consumer behaviour influences not only sales but also production. Consumers thus become an important incentive for product innovation and production of new ones.

Today it is impossible to imagine a life without shopping. Shopping has become our common activity. Shopping behaviour is a significant sociological phenomenon, understood as a manifestation of a consumer way of life, largely influenced by a society in which the consumer uses goods and services. It includes reasons that lead the customer to become a consumer of a particular product. It is all the more difficult to anticipate the shopping behaviour of the silver generation because demographic, social and economic changes will continue in Europe [38]. These trends are also very important for the Slovak economy and society and will also have to pay increased attention.

Every retailer needs to know their customers to develop their business. It obtains information from several sources, e.g. from loyalty cards and monitoring of shopping movements in shopping centres, but mainly from various surveys, through which it obtains a more comprehensive picture of the customer. Behavioural surveys make it possible to obtain the so-called soft data. The data obtained from the field survey conducted in Nitra in 2015 and in Trnava in 2014 were used to analyse the shopping behaviour of the silver generation.

By the silver generation we mean the population of post-production age over 50 years, which we divided internally into two groups:

1. Group 50 and 60 years old (younger seniors)
2. Group 60+ (older seniors)

In the city of Nitra, 217 respondents were addressed, of which 109 were in the age category 50–60 years old and 108 in the category 60 years old and older, and in Trnava 220 respondents, of which 111 were younger and 109 older seniors. The questionnaire was divided into three basic parts. The first part was focused on knowledge of the respondent's demographic characteristics (gender, age, educational attainment, occupation, permanent residence). In the second part of the questionnaire respondents determine the factors by selected domains that affect them when shopping. In the third part, at their own discretion, they were included in different types of consumer behaviour, based on the methodology of the company [39]. Trembošová et al. [40], based on the form, manner and motivation of shopping, established Shopper Typology Media Behaviour—two basic types of shopping orientations: first modern and second traditional.

1. Modern shopping orientation consists of three shopping subtypes of customers:
   - Influential: Emotional in shopping behaviour, influenced by advertising, purchases and stock items, makes a large number of purchases in one day.
   - Demanding: Has high demands on the quality of goods and shopping comfort.
   - Mobile (pragmatic): Optimizes the ratio between price and value of goods, prefers large-scale stores and regularly uses the car.
2. The traditional shopping orientation consists of four shopping subtypes:
   - Cautious (conservative): Makes rational and conservative decisions, does not trust advertising, does not use a car, has a low share of impulsive purchases.
   - Saving: Minimizes costs, purchases rationally, uses little car.
   - Loyal: Purchases frequently and in small quantities, prefers a small shop near the home, professes tradition, focuses shopping on the social side of life.
   - Unpretentious (phlegmatic): He/she has no demands on the shop, prices are indifferent to him/her, he/she does not travel for shopping.

Respondents were approached outside shopping centres to avoid being directly involved influence. The polls were mainly streets, pedestrian zones and bus stops. Consumer respondents' preferences and opinions were evaluated by classical method—comparative analysis based on description, as well as statistical, graphical and cartographic methods.

For smaller households, seniors shop more often but make smaller purchases, especially food, fruit and vegetables.

4 Analysis and Results

According to the Eurostat forecast [41], the proportion of the population will gradually decrease in productive age (20–64 years), especially after a generation born in a period of increased birth, i.e. after World War II, and it will gradually reach retirement age (the strongest pressure in this direction is expected in the period 2015–2035). According to InfoStat forecasts, seniors will make up about 30% of the total population in Slovakia in 2050. It is assumed that the Slovak population will be one of the fastest ageing in the whole European Union. According to demographic forecasts, Slovakia will change in the short term from one of the youngest to one of the oldest countries in the EU (basis of preparation of national priorities of the Slovak Republic in 2017). In 2030 there will be 1.38 workers per pensioner, but in 2060 it will be only 0.98 workers and Slovakia will increasingly face a labour shortage. For the needs of Slovakia, the "National Program of Active Aging for 2014–2020" has been prepared, which focuses primarily on the issues of employment and unemployment and less on the long-term sustainability of the workforce stock, in which senior citizens will have a significant position.

The ageing process is characteristic of not only rural areas but also Slovak cities. In addition to the cities of Bratislava and Košice, cities are experiencing a decline in the population, caused by a negative natural increase, but also by a low migration balance. The regional towns of Nitra and Trnava also belong to the group of "shrinking" of towns [42], in which the population is rapidly decreasing. Population outflow strengthens the suburbanization process, which enhances the growth of their background. According to Haase et al. [43] the decline in urban population is a common phenomenon in the post-socialist region. A decrease in the population is also expected in regional cities of western Slovakia, including Nitra and Trnava. Nitra ranks sixth in the number of inhabitants after Bratislava, Košice, Prešov, Žilina, Banská Bystrica and Trnava [7]. As the population of cities decreases, the proportion of the population of the silver generation in the city's population is increasing. In 2035 it should reach 46.04% in Nitra and 46.39% in Trnava [7].

Both cities are among the oldest cities in Slovakia, with a rich commercial tradition. The oldest historical reports about Nitra come from the 30th of the 9th century from the Pribina's period. They relate to a castle whose existence is confirmed by the construction of the first Christian church in Slovakia. It was located under the castle, a market settlement that also served a commercial function. Trnava is known as the oldest royal town of Slovakia, which was established at the crossroads of old trade routes as a merchant settlement. The first written mention of it dates back to 1211. In the Middle Ages, the important trade route Via Bohemia, connecting Prague–Brno–Hododín through Trnava with Esztergom and Buda, passed through the town. On this significant communication the city of Nitra was also connected to the road. Both cities still benefit from excellent geographic location. They are part of the development axis of Slovakia's settlement Bratislava–Banská Bystrica–Košice. They have a supra-regional position in the communication

Table 1 Basic indicators of retail network of regional cities Nitra and Trnava in 2015

Indicator	Nitra	Trnava
Number of stores	871	769
Number of stores per 1000 inhabitants	10.87	11.93
Sales area	160,919	120,559
Average store area	184.7	157.0
Area parameter m^2/1000 inhabitants (PAFS[a])	2186	1870
Operating parameter number of inhabitants/1 shop (PS[b])	92.6	83.8
Number of shopping centres	4	5
Number of stores in shopping centres	347	206
Gross leasable area	100,409	83,072
Number of large stores (VP[c] over 2500 m^2)	19	25
The population	**80,161**	**64,439**

Source: Authors' survey
[a]PAFS the ratio of the population (in thousands) to the admissible floor space
[b]PS the ratio of the population (individual) and number of stores
[c]Large-scale stores

and settlement structure of western Slovakia. Nitra and Trnava as regional cities belong to the category of settlements with the highest standard of tertiary service. They are located on the R1 expressway. Good transport accessibility has become a major factor in the foreign capital inflow of multinationals into the economy. Cities are known mainly by the automotive industry, which increases the capability of their population to other regional cities Trenčín, Banská Bystrica and Prešov. Despite the fact that both evaluated cities have the same position in the settlement structure of Slovakia, their development after the change of political-economic conditions (after 1989) was different, which was also reflected in the development of retail network—in the number of stores, in the size of the sales area, etc. (see Table 1). In terms of the number of stores, Nitra has a primary position, which exceeds Trnava. Field research in the city of Nitra identified 871 retail units (of which 19 are large-scale stores such as department store, supermarket and hypermarket). In Trnava, the retail network consists of 769 retail units (of which 25 are large retail stores). A similar situation is in the sales area. The larger sales area in Nitra reflects the greater representation of large-scale stores over 2500 m^2 in the outskirts of the city and the reduction of smaller stores in the shopping area in the central part of the city.

Between 2001 and 2015, 9 shopping centres were built in the surveyed cities (of which in Nitra 4 and Trnava 5), which have a total gross leasable area (HPP) of 100,409 m^2 (see Table 1). According to the location, three of them NC Centro, Max and Gallery belong to the type, according to Guy [44] classification, out-of-centre and NC Mills to edge-of-centre in Nitra. In Trnava, there are five shopping centres with a total gross rentable area (HPP) of 83,072 m^2 (Table 2). Of these, three are out-of-centre. The newest shopping centre of Trnava OC City Arena (opened in 2015), which is unique in Central European space in connection with a football stadium, is

Table 2 Basic indicators of shopping centres in Nitra and Trnava in 2015

City	Name	City Boroughs name	Opening year	Number of stores	HPP[a] in m^2	Number of parking place
Nitra	OC Centro	Chrenová	2006	90	23,982	764
	ZOC Max	Chrenová	2006	75	16,487	450
	OC Galéria	Staré Mesto	2008	43	27,440	686
	OC Mlyny	Staré Mesto	2009	139	32,500	1100
	Mesto Nitra			47	100,409	3000
Trnava	ZOC Max	Trnava-Stred	2004	37	10,000	126
	OC Galéria	Trnava-Sever	2006	43	15,072	845
	SP Arkadia	Trnava-Sever	2006	28	8000	1285
	The Mall Trnava	Trnava-Juh	2008	30	30,000	
	OC City Aréna	Trnava-Stred	2015	68	2000	950
	Mesto Trnava			206	83,072	3206

Sources: [40], authors' survey
[a]HPP = gross leasable area

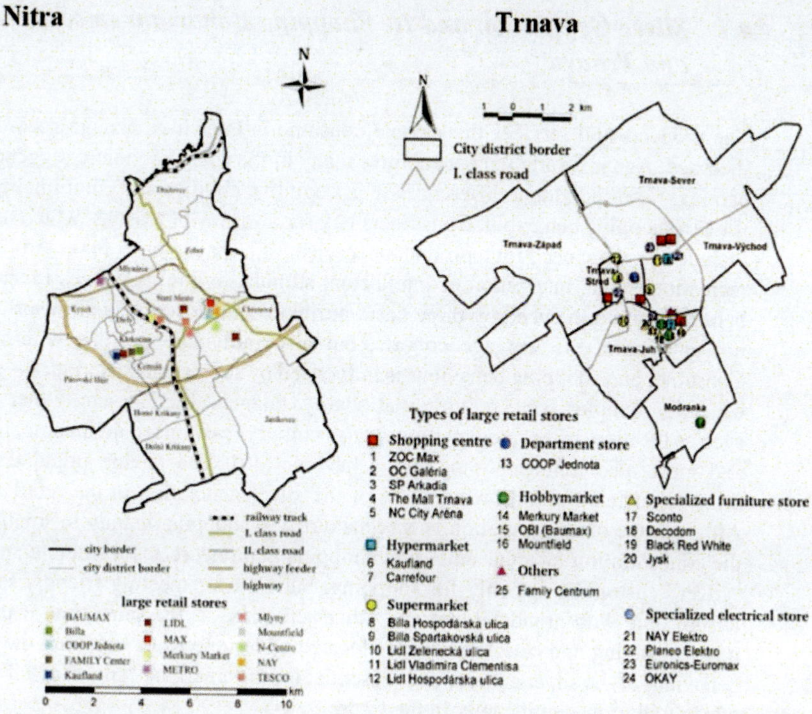

Fig. 1 Spatial distribution of retail large-scale stores in Nitra and Trnava in 2015. Sources: [40], authors' survey

included in the "edge-of-centre" type. The centres Max, Gallery and Arkadia belong to the out-of-centre type and, depending on the spatial character, it is a mixed type 1 and 2 sites characterized by spatial proximity to customers' homes and competing companies. The Mall Trnava belongs to the group of shopping centres "edge-of-town" (see Fig. 1).

Both cities as centres of the automotive industry are significant centres of commuting from the territory of western Slovakia and also centres of trade. The most significant is the daily commuting from their background. Inhabitants who commute to work use their services, healthcare, education and others, while at the same time making purchases of daily and durable goods, increasing the pressure on retail services, especially on shopping centres near bus and railway stations.

4.1 Silver Generation and Its Shopping Behaviour in Nitra and Trnava

The products and services that people consume in their lives change according to their age. Age is a variable that captures many of the socio-economic (eggeneration earnings) and individual differences (e.g. cognitive abilities and emotionality) also used in shopping behaviour. Consumers buy for a variety of reasons, which may not include a specific need for a product or service [45], for example, for entertainment, recreation, social interaction or intellectual stimulation [46]. Changes in shopping behaviour are influenced by three basic attributes, need, supply and means, which not only have different age cohorts, but also individual persons within them. Consumer and shopping behaviour is influenced by several characteristics—gender, education, employment, religion, nationality, financial income, etc. Other factors such as location, price, product range, sales culture, real household income, lifestyle and more also influence shopping behaviour [40]. The ageing population also brings changes in the growing share of the silver generation in the retail market. Although the older generation was accustomed to shopping mainly in small stores, the concentration of retail units into shopping centres creates a suitable space for "silver" customers not only for shopping, but also for meeting friends, spending leisure time with grandchildren and other activities. At the same time it provides postal, banking and catering services located under one roof. Knowing the buying behaviour of the silver generation is essential for the emerging "silver" retail market, which will grow significantly in the future.

The silver generation group consisted of 437 respondents over 50 years. Both cities are represented by almost balanced share in the number of respondents, Nitra 49.66%:Trnava 50.34%. The segment of younger seniors consisted of 220 respondents, of which Nitra had 109 and Trnava 111. The segment of older seniors was formed by respondents over 60 years old. This group consisted of 217 respondents of which 108 were in Nitra and 109 in Trnava. Among the seniors women have a more prominent position in both cities. Women achieved 57.67% of the sample studied. It is generally known that women buy rather than men, which has been reflected in this generation. There were 132 senior women per 1 senior buyer. The ratio of respondents in both cities was almost balanced (Nitra 128:Trnava 124), as in both categories of seniors (see Fig. 2a). Nitra and Trnava are old centres of the secondary and higher (university) education, which was also reflected in the structure of seniors according to the achieved education. Most respondents (77.80%) have completed secondary education (see Fig. 2b).

Higher education is represented in 8.24% and primary education in 13.96% of respondents. The silver generation is specific to the job structure (see Fig. 2c). Segment 50–60-year-old, younger seniors are involved in work, are in the process of preparing for retirement, of which 15.9% are retired, especially premature. The age group 60 and over is made up mainly of pensioners (93.55%); however, working pensioners (6.45%) are also represented in the group. Basic structures of respondents in both cities by age cohorts are documented in Fig. 2.

a) Structure of respondents by the gender

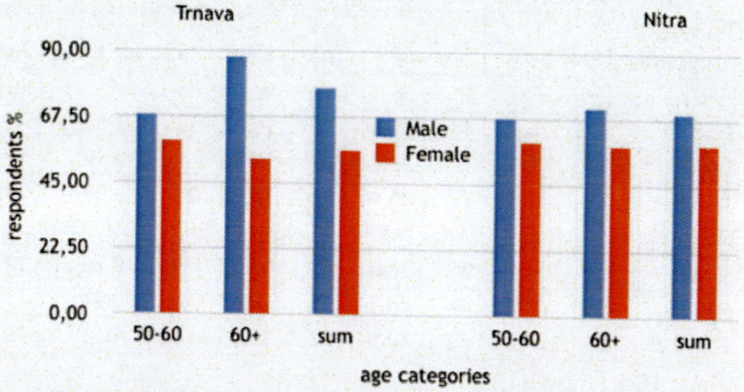

b) Structure of respondents according to educational attainment

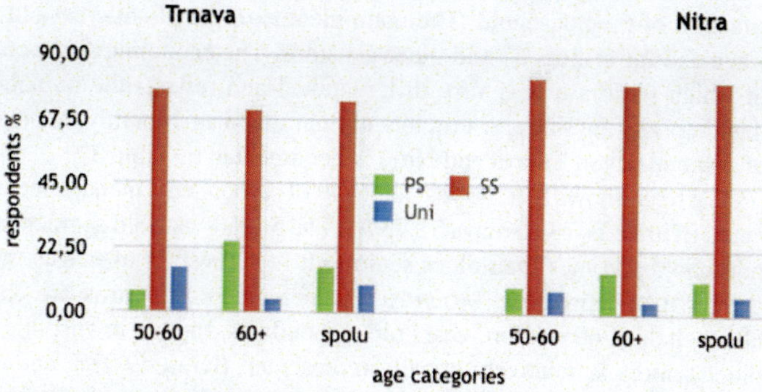

c) Employment structure

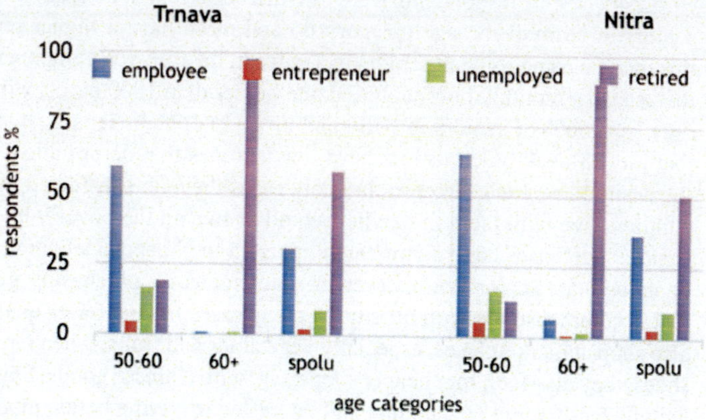

Fig. 2 Basic characteristics of respondents in Nitra and Trnava. (**a**) Structure of respondents by the gender. (**b**) Structure of respondents according to educational attainment. (**c**) Employment structure. Source: Own research

Table 3 Monthly household income of respondents in Nitra and Trnava

Town/age categories	Monthly household income (euro)										
	Up to 500		501–1000		1001–1500		1501 and more		Not specified		Together
Trnava	abs.	%	abs.	%	abs.	%	abs.	%	abs.	%	abs.
50–60	27	24.32	44	39.64	17	15.32	15	13.51	8	7.21	111
60+	54	49.54	35	32.11	15	13.76	5	4.59	0	0.00	109
Together	81	36.82	79	35.91	32	14.55	20	9.09	8	3.64	220
Nitra											
50–60	10	9.17	29	26.61	47	43.12	9	8.26	14	2.84	109
60+	53	49.07	44	40.74	5	4.63	1	0.93	5	4.63	108
Together	63	29.03	73	33.64	52	23.96	10	4.61	19	8.76	217

Source: Own research

The shopping power of seniors depends mainly on income and economic situation of the household. The main income of the 50–60-year-old segment is the wage and for seniors 60 and older pensions. The economic situation of households in which seniors live is very differentiated and reflects the number of household members and the size and structure of their incomes. Monthly income of households of respondents in Trnava and Nitra is documented in Table 3.

The table shows that the vast majority of respondents manage incomes up to 1000 euros (Nitra 62.67%, Trnava 72.73%). The 50–60-year-old segment achieves higher household incomes than older seniors. It is a question of wages, which is always greater than retirement. Monthly incomes up to 500 euros are characteristic for almost half of 60 or more years old respondents. In both towns, this income among older seniors is relatively equally represented (Nitra 49.07% and Trnava 49.54% of respondents). The growth in monthly household incomes decreases the number of respondents. In households with older seniors with incomes from 500 to 1000 euros, 40.74% of respondents were in Nitra and 32.11% in Trnava. It follows those seniors in Nitra at the age category 60 and more have a higher representation in this income category than seniors in Trnava in the research component. Evidence of this is also a monthly household income higher than 1000 euros, which is managed only by 5.56% of seniors in Nitra and up to 18.35% in Trnava. It should be noted that most Slovak older seniors have low incomes that do not allow them to live at the desired level and therefore in many cases they are subsidized by their working children, live with them in one household or live on the poverty line. The range of seniors' purchases has a significant impact on the financial situation of a household or individual. Seniors with lower incomes focus on purchasing goods of a price that is commensurate with quality, in many cases buying fewer quality goods; they also shop at marketplaces, especially vegetables and fruits; when buying clothes and shoes, they also look for cheaper shops, e.g. with Chinese goods. Solvent consumers are inherently more demanding, and interested in buying better, more luxurious and fashionable products.

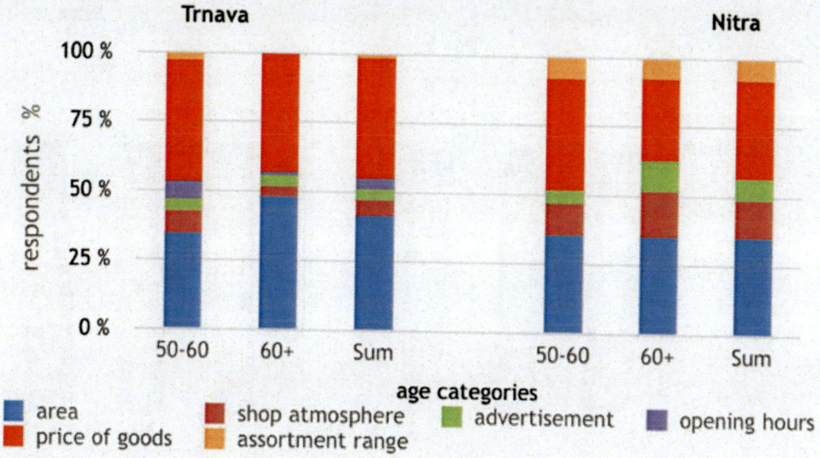

Fig. 3 Factors affecting purchasing. Source: Own research

The attractiveness of shopping units is enhanced by their location, presented mainly by convenient accessibility by means of transport. It is one of the key factors affecting customers [47]. Seniors often visit edge-of-centre.

Out-of-centre places are easily accessible by public transport and provide ample parking space. For everyday shopping, the majority of seniors prefer to shop in retail units closest to their homes. In the case of employed seniors, retail establishments located in the vicinity of the workplace are mainly focused on selling food. The price of goods is one of the decisive factors in choosing a retail store alongside the location of a retail facility. They are also a primary factor for the seniors of both studied urban segments. Internal response structure of seniors is differentiated. The price of younger seniors is higher when shopping in both cities (Nitra 43.88%:Trnava 35.53%, Nitra 40.22%:Trnava 35.33%), while for older ones the location and its accessibility by public transport (Trnava 48.15%:Nitra 42.96%; Nitra 34.97%:Trnava 29.37%); see Fig. 3.

The price of goods is closely related to various sales promotions that carry out retail operations to attract the customer. The reduced price of products is monitored by seniors. Especially the younger group of seniors are focused mainly on branded products. Information about stock prices is obtained by seniors through ads and leaflets that reach the household. Advertising is paid more attention by Nitra respondents (7.40%) than Trnava respondents (4.01%). Shopping centres are characterized not only by the range of retail operations, but also by the range of products in them. They allow you to buy a wide range of products under one roof. There are several types of branded stores, mainly with textile and clothing products, which present mainly assortment and fashion trends, which are mainly focused on the younger segment of respondents. Older group of seniors favour sales promotions and product range, especially in food products. The use of reduced prices for food

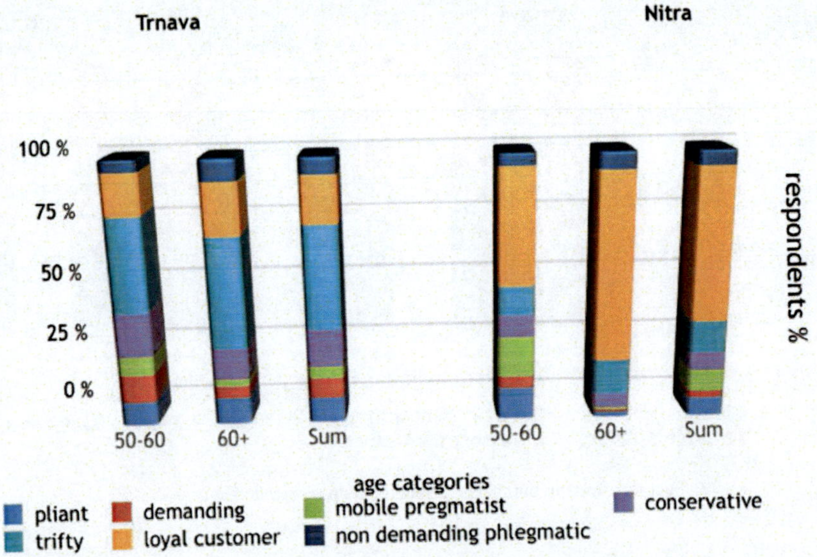

Fig. 4 Structure of types of shopping behaviour of seniors in Nitra and Trnava. Source: Own research

or other products is mainly a reflection of the financial situation of older people, which is often very complicated.

The purchase is also determined by the atmosphere of the store, consisting mainly of the speed of service and its willingness, the size of the store, and also the arrangement of the goods. It is the shop climate that leads older people to prefer certain stores to others. The respondents treat them loyally and often visit them. This factor is particularly favoured by Nitra seniors.

According to Spilková [48], buying behaviour is defined as the behaviour that consumers show when searching, buying, using ratings and using products and services from which they expect to meet their needs. It includes the reasons that lead the customer to become a consumer of a particular product. Consumer behaviour is influenced by the availability of goods, culture, sales, selling concepts, real household income, lifestyle and retailer marketing activities. In addition to the above-analysed attributes, the behaviour of seniors is influenced by a number of other objective and subjective factors; therefore, respondents were given the option to be classified in one of the seven subtypes. As seen in Fig. 4, seniors were included in all subtypes. According to the nature of the shopping subtypes, we can classify them into two basic groups. The first group consists of subtypes, inclined to the modern way of shopping. These include mobile, demanding and influenced. The second group consists of subtypes leading to the traditional type of shopping (cautious, saving, loyal and unpretentious).

The older generation tends to the traditional type of shopping, which shows a significant share of savings and loyalty, which is also reflected in the seniors of

Nitra and Trnava. 82.03% of respondents were included in the traditional type in Nitra and 78.18% in Trnava. More than one-fifth of respondents have integrated into a modern type of shopping behaviour. They were 21.82% in Trnava and 17.97% in Nitra. In both cities the modern type of shopping behaviour is also represented by all its subtypes—influential, demanding and mobile pragmatist. The modern type is more pronounced in the group of younger seniors, 31.11% of respondents in Nitra and 26.76% in Trnava.

In the traditional shopping type of seniors Nitra prevailed a loyal subtype, which included more than half (57.60%) of the respondents. The highest values were expected in both groups of seniors. The subtype is characterized by frequent buying in small quantities. Respondents prefer shopping in stores near their homes. The group of respondents has a more prominent position in the group of respondents over 60 years, it exceeds (1.5 times) the share of the group of 50–60 years. The younger group of seniors in this subtype reaches 44.04%. The traditional type also includes a subtle type of savoury and unpretentious phlegmatic. Saving customer is characterized by minimized spending and rational purchasing, especially the goods he/she needs. The group of younger seniors represents 11.98%. In the group of older seniors, it reaches a higher value—12.96%. Other subtypes reported by seniors were conservative customer (6.91%) and unpretentious phlegmatic (5.53%). Shopping behaviour of a conservative (cautious) customer is typically rational and conservative decision-making, does not trust advertising and is characterized by a low proportion of impulsive purchases. This subtype has a higher proportion in the group of younger seniors (8.26%) and in the group of older seniors it reaches 5.56%. The customer of the unpretentious phlegmatic subtype is characterized by the fact that he/she has no demands on the shop, prices are indifferent, he/she does not travel for purchases and he/she predominates in the category of older respondents (6.48%). As seen in Fig. 4 the group of 60 and more years have a significant position in just two subtypes, loyal and saving, while the other subtypes are at least represented. In the group of 50–60-year-old respondents, almost one-third (32.11%) belongs to the modern type, which is also related to higher incomes in respondents' households in the city of Nitra. Especially working women in the way of life and especially in fashion do not want to lag behind women of younger ages. Within this type, the mobile pragmatist (15.60%) and the influential customer (11.93%) have a higher position. Influenced customer is emotional in shopping behaviour, is influenced by advertising, makes more purchases within one day and purchases branded goods. The mobile pragmatist optimizes the price/value ratio of the goods and prefers to buy in large stores.

Unlike in Nitra, the seniors of Trnava have a stronger representation in the savings subtype. Respondents of this subtype are characterized by minimization of expenditures; they purchase rational, mostly only necessary, items. They represent the dominant share of both groups, where they reach 35.71% in the younger segment and a 42.9% share in the older segment of seniors. In second place is the loyal subtype, represented by a value lower by half than the saving subtype (18.18%). This subtype is more prevalent among respondents over 60 (20.18%) than among younger respondents of 50–60 years old (16.7%). The traditional type also includes a more conservative customer, which is more pronounced in the group of 50–60 years (16.96%) than in the group of seniors older than 60 years (11.93%). The modern type of Trnava seniors slightly exceeds Nitra seniors (21.82:17.97). This type of shopping behaviour of seniors is more strongly represented in its subtypes. In the subtype of influential customers 9.55% of seniors in Trnava were included; the older segment of seniors had a higher share (10.09%) and younger represented 8.93%. As expected, the other two subtypes of demanding and mobile pragmatists are characterized by a higher percentage in the 50–60-year-old group.

The demanding subtype is almost three times higher than this subtype in Nitra. The opposite situation is in the subtype of a mobile pragmatist, which in turn is twice as high in the group of Trnava respondents. The representation of individual subtypes in the research component of Nitra and Trnava seniors is documented in Fig. 3. The survey shows that seniors tend mainly to traditional shopping, but on the other hand, modern shopping also has a significant share (80.09:19.91). This share reflects mainly the younger segment of seniors who do not want to lag behind the younger generations in their way of life. While the mobile pragmatist has a significant presence in the modern type, which is subsequently influenced and demanding customer, the traditional type is dominated by a loyal, saving and unpretentious customer.

5 Conclusions

Population ageing is a characteristic demographic process of the current period, which will deepen in the future. Generation 50 years and over, known as the silver generation, with an increasing labour shortage, has become an increasingly important economic and social force whose position in society is growing. This generation, not homogeneous, is intrinsically diverse not only in terms of age but also in terms of its characteristics and needs. Already today they form a strong group of consumers who are not given adequate attention. There is still a big gap in the consumer market between what older consumers want and need and what the market offers them. It is therefore necessary to pay more attention to their demands, requirements and behaviour, including the retail sector. The younger segment of the silver generation (50–60 years old), represented mainly by women, is trying to "rejuvenate" to imitate some behaviour (e.g. dressing, leisure) of the younger generation, which is also directly reflected in their shopping behaviour. This decision-making process for older consumers provides the basis for the development of the mega-trend known in the literature as a "rejuvenating population". It is a modern type of buying behaviour with subtypes represented by a mobile pragmatist, an influential and demanding customer.

In addition to shopping, seniors will increasingly use the wide range of services provided by shopping centres in relation to changes and new trends in leisure behaviour, e.g. sitting with friends over a coffee, spending free time with grandchildren in children's corners, walking, strengthening health in fitness centres and visiting cinemas.

In Slovakia, older seniors (60 years and over) achieve lower incomes than seniors in Western Europe; in many cases they depend on economically active children who contribute to the care of their parents. Most of the respondents of the silver generation in the regional towns of Nitra and Trnava live in households with monthly income up to 1000 euros (Nitra 62.67%, Trnava 72.73%). Respondents' households of up to 500 euros have a significant position among them. The growth in monthly household incomes decreases the number of older respondents. Monthly household incomes over 1000 euros are represented in only one-quarter of respondents' households (Nitra 28.57% and Trnava 23.64%). Very low monthly incomes up to 500 euros are shown mainly by households of older seniors aged 60 and over. Seniors with lower incomes buy cheaper and lower quality goods; they orient themselves in the market, where there are cheaper, but fresh products, especially fruits and vegetables. They search for cheaper stores, e.g. with Chinese goods. Solvent consumers are inherently more demanding, interested in buying better, more luxurious and fashionable products. When shopping, seniors prefer well-accessible retail facilities, especially in the central parts of the city. For daily shopping, they prefer shopping in retail units as close to their home as possible. Employed seniors are sought-after retail establishments located near the workplace focused mainly on the sale of food. The decisive factor is also the price of the goods, which is associated with various promotions of retail facilities to attract customers. In the younger group of seniors reduced price is interesting for buying branded products. Older seniors are interested in promotions and the range of products mainly oriented towards food products. Older seniors also prefer shopping in stores with a familiar atmosphere, which is created by the speed of service and its willingness, the size of the store, the arrangement of goods, etc. They treat them loyally and often visit them. Seniors tend predominantly to the traditional type of shopping with a smaller proportion of the modern type. This share reflects the interest of the younger segment of seniors in particular, who do not want to lag behind in their way of life, especially dressing up for younger generations. While the mobile pragmatist has a strong presence in the modern type, which is subsequently an influenced and demanding customer, the traditional type is dominated by loyal, saving and unpretentious, which is characteristic for seniors of both cities.

As income and living standards grow, modern shopping behaviour will grow. An important segment is online shopping, which in this part of the population has a strong position and has risen sharply due to the COVID-19 pandemic. We assume that shopping will move mainly to online space, which will be the subject of further research. We believe that seniors will reflect on the new situation in shopping and restricting shopping in brick-and-mortar stores by using new forms of online communication. It is important for the retail network to know the ways of shopping not only by age but also by gender and especially by the range of goods, with emphasis on different spatial scales. In the post-pandemic period, it is important to point out the changes in spatial and assortment structures in retail caused by the influence of COVID-19.

Acknowledgements This research was supported and funded by 030STU-4/2018 KEGA project titled "E-platform as basis for improving collaboration among universities and industrial enterprises in the area of education", and supported by 2/0077/19 VEGA project titled "Work competencies in the context of Industry 4.0", project APVV-18-0185 Land-use changes of Slovak cultural landscape and prediction of its further development, project 1/0880/21 VEGA Transformation of the Nitra Region in changing socio-economic conditions with special regard to the effects of the COVID-19 pandemic and project HO2020Nr. 873134 "Linking Research and Innovation for Gender Equality".

References

1. P. Alders, D.P. Broer, Ageing, fertility and growth. J. Public Econ. **89**(5–6), 1075–1095 (2005)
2. D.J. Van De Kaa, The idea of a second demographic transition in industrialized countries. Jpn. J. Popul. **1**(1), 1–34 (2003)
3. D. Cagáňová, *Multiculturality and Industrial Enterprises*, 1st edn. (Hochschule Anhalt, Köthen, 2011)
4. T. Zalega, Smart shopping—nowy trend konsumencki [Smart shopping: A new consumer trend]. Handel Wewnętrzny **6**(347), 43–55 (2013)
5. The 2018 Ageing Report 4: Economic and Budgetary Projections for the EU Member States (2016–2070). https://ec.europa.eu/info/publications/economic-and-financial-affairs-publications_en. Accessed 11 May 2021
6. The 2030 Agenda for Sustainable Development 5: Available at https://sustainable development.un.org/content/documents/. Accessed 11 May 2021
7. B. Šprocha, B. Bleha, B. Vaňo, J. Buček, *Perspektívy, Riziká a Výzvy Demografického Vývoja Najväčších Miest Slovenska*, 1st edn. (Univerzita Komenského, Bratislava, 2017)
8. L.A. Morgan, S.R. Kunkel, *Ageing Society and Life Course*, 4th edn. (Springer, New York, 2011)
9. T. Zalega, Smart shopping in consumer behaviour of Polish seniors (Report from own research). Acta Sci. Pol. Oeconomia **16**(3), 101–110 (2017)
10. L.P. Fried, W.J. Hall, Leading on behalf of an ageing society. J. Am. Geriatr. Soc. **56**(10), 1791–1795 (2008)
11. T. Zalega, Konsumpcja osób starszych w Polsce [Consumption among Elderly People in Poland]. Nierówności Społeczne a Wzrost Gospodarczy [Social Inequalities and Economic Growth]. Zeszyt **42**(2), 152–173 (2015)
12. S. Biggs, C. Phillipson, R. Leach, A. Money, Culture of consumption. Baby boomers and adult aging in public policy: the changing relationship between production and consumption. Working Paper 27 (2006)
13. I. Pauhofová, M. Páleník, Súvislosti starnutia európskej populácie vo väzbe na formovanie dopytu seniorov. Working papers, 37 (2012)
14. B. Ying, R. Yao, Consumption patterns of Chinese elders: evidence from a survey in Wuhan, China. J. Fam. Econ. Iss. **27**(4), 702–714 (2006)
15. Ľ. Nagyová, J. Stávková, Z. Kádeková, Income disparities and their impact on level and development of food expenditures of households in Slovakia. Acta Univ. Agri. Silvic. Medelianae Brun. **61**(7), 2569–2575 (2013)
16. A. Singh, L.M. Kathuria, Understanding drivers of branded food choice among low-income consumers. Food Qual. Prefer. **52**, 52–61 (2016)
17. B. Casey, Dôsledky hospodárskej krízy pre dôchodky a dôchodkovú politiku v Európe. Globálna Sociálna Politika **12**, 246–265 (2017)
18. World report on ageing and health. https://www.who.int/ageing/en/. Accessed 4 June 2021
19. C. Schaffnit-CHatterjee, Wie werden ältere Deutsche ihr Geld ausgeben? Aktuelle Themen 385, Deusche Bank Reasearch (2007)

20. A. Kasčaková, L'. Kubisová, Social and economic potential of silver population in Slovakia. Prace Naukowe Uniwersytetu Ekonomicznego we Wroclawiu **435**, 125–135 (2016)
21. V. Páleník et al., *Strieborná Ekonomika v Slovenskom, Európskom a Svetovom Kontexte*, 1st edn. (Ekonomický ústav, Bratislava, 2012)
22. R. Leach, S. Biggs, C. Phillipson, A. Money, Boomers and Beyond: intergenerational consumption and the mature imagination. http://www.consume.bbk.ac.uk/research/biggs. html#outline. Accessed 4 June 2021
23. J.F. Coughlin, Lecture on Aging, Innovation and Policy. Massachusetts Institute of Technology. http://www.cscollege.gov.sg/cgl/events/
24. T. Olejniczak, Senior consumer behaviour-worldwide research areas census polish ones. Acta Oeconomia **15**(4), 127–135 (2016)
25. R.C. Leventhal, Aging consumers and their effects on the marketplace. J. Consum. Mark. **14**(4), 276–281 (1997)
26. S.A. Nasco, D. Hale, Information search for home, medical, and financial services by mature consumers. J. Serv. Mark. **23**(4), 226–235 (2009)
27. S. Pettigrew, K. Mizerski, R. Donovan, The three big issues for older supermarket shoppers. J. Consum. Mark. **22**(6), 306–312 (2005)
28. N.J. Miller, K. Soyoung, The importance of older consumers to small business survival: evidence from Rural Iowa. J. Small Bus. Manag. **37**(4), 1 (1999)
29. A.J. Greco, L.E. Swayne, Sales response of elderly consumers to point-of-purchase advertising. J. Advert. Res. **32**(5), 43 (1992)
30. Carstensen et al., Motivation for Social Contact Across the Life Span: A Theory of Socioemotional Selectivity, in *Nebraska Symposium on Motivation*, Developmental Perspectives on Motivation, ed. by J. Jacobs, vol. 40, (University of Nebraska Press, Lincoln, 2003), pp. 209–254
31. P. Williams, A. Drolet, Age-related differences in responses to emotional advertisements. J. Consum. Res. **32**(3), 343–354 (2005)
32. D. Cagáňová, A. Stareček, N. Horňáková, P. Ridzoňová Hlásniková, The analysis of the Slovak citizens' awareness about the smart city concept. ACM Mob. Netw. Appl. **24**(6), 2050–2058 (2019)
33. N. Vraňáková, K. Koltnerová, A. Chlpeková, D. Cagáňová, Application of exact methods in employee selection in accordance with the age management concept. ACM Mob. Netw. Appl. **24**(6), 2038–2042 (2019)
34. R. Lambert-Pandraud, G. Laurent, Tell me which perfume you wear, I'll tell how old you are: Modeling the impact of consumer age on product choice. Working paper. HEC Paris (2007)
35. R. Lambert-Pandraud, G. Laurent, E. Lapersonne, Repeat purchasing of new automobiles by older consumers: empirical evidence and interpretations. J. Mark. **69**(2), 97–113 (2005)
36. Cuddy et al., Do managers' perceptions coincide with established stereotyping of older workers' capabilities? Psychol. of Stereotypes. **7**, 173–192 (2005)
37. M. Rypáková, Význam sledovania trendov v procese nákupného správania spotrebitel'ov v podmienkach globalizácie. In: Medzinárodná vedecká konferencia Globalizácia a jej sociálno-ekonomické dôsledky (2013). https://www.researchgate.net/publication/260019509_Vyznam_sledovania_trendov_v_procese_nakupneho_spravania_spotrebitelov_v_podmienkach_globalizacie
38. G. Desvaux, B. Regout, *Meeting the 2030 French Consumer. How Europeanwide Trends Willshape the Consumer Landscape* (McKinsey & Company)
39. V. Stárová, Segmentace nákupního chování—trend i příležitost. Výsledky Shopper Typology & Media Behaviour, Incoma Research a Gfk Praha (2003), https://www.kurzy.cz/zpravy/62763-segmentace-nakupniho-chovani-trend-i-prilezitost/. Accessed 10 June 2021
40. M. Trembošová, A. Dubcová, H. Kramáreková, L'. Nagyová, *Nákupné Správanie Obyvatel'stva Mesta Nitry - Vývoj a Zmeny*, 1st edn. (SPU, Nitra, 2016)
41. European Commission Demographic Report. https://op.europa.eu/en/publication-detail/-/publication/615463f7-9e58-45a8-adbf-81ba1ea88d7d/language-en. Accessed 10 June 2021

42. J. Buček, B. Bleha, Urban shrinkage as a challenge to local development planning in Slovakia. Morav. Geogr. Rep. **21**(1), 2–15 (2013)
43. A. Haase, M. Bernt, K. Grossmann, V. Mykhnenko, D. Rink, Varieties of shrinkage in European cities. Eur. Urban Reg. Stud. **23**, 86–102 (2016)
44. C. Guy, Classifications of retail stores and shopping centres: some methodological issues. Geo J. **45**(4), 255–264 (1998)
45. E.M. Tauber, Marketing notes and communications: why do people shop? J. Mark. **36**(4), 46–49 (1972)
46. M.J. Arnold, K.E. Reynolds, Hedonic shopping motivations. J. Retail. **79**(2), 259–268 (2003)
47. L. Nagyová, Zmeny na slovenskom trhu potravín v dôsledku vstupu zahraničných obchodných sietí. Ekonomika firiem. Bratislava: Ekonomická univerzita (2000), pp. 451–456
48. J. Spilková, *Geografie Maloobchodu a Spotřeby—Věda o Nakupování*, 1st edn. (Karolinum, Praha, 2012)

Impacts of the Covid-19 Pandemic on the Educational Process: Use of Innovative Educational Technologies

Jana Smetanková, Peter Mésároš, Katarína Krajníková, and Marcel Behún

1 Introduction

The Covid-19 pandemic is a global health crisis and the biggest crisis since the World War II. The virus began to spread from Asia in late 2019 and affected all continents. The pandemic affected all aspects of society. There are currently nearly 150 million infections and approximately 3.2 million deaths from Covid-19. Pandemics pose a high health risk and cause a socioeconomic crisis. People lose their jobs and live in constant insecurity [1].

Due to the pandemic situation, individual employers were forced to implement key measures to protect against Covid-19 in the workplace. The World Health Organization has recommended raising occupational hygiene standards by increasing the frequency of handwashing or disinfecting hands with alcohol-based disinfectants and increasing hygiene—air purification, physical distance support (at least 1 m), wearing face masks and respirators, regular cleaning and disinfection of common areas, and preventing the movement of employees [2, 3].

The pandemic affected all sectors of society. Individual educational institutions had to move from full-time education to distance learning. Many institutions were closed or operated in a partially closed regime. As many as 91% of students were affected by the temporary closure of schools.

Most of the governments around the world have temporarily closed educational institutions to limit the spread of the pandemic. For example, in China, the beginning

J. Smetanková (✉) · P. Mésároš · K. Krajníková
Faculty of Civil Engineering, Technical University of Košice, Košice, Slovak Republic
e-mail: jana.smetankova@tuke.sk; peter.mesaros@tuke.sk; katarina.krajnikova@tuke.sk

M. Behún
Institute of Earth Resources, Technical University of Košice, Košice, Slovak Republic
e-mail: marcel.behun@tuke.sk

L. Knapčíková, D. Peraković (eds.), *6th EAI International Conference on Management of Manufacturing Systems*, EAI/Springer Innovations in Communication and Computing, https://doi.org/10.1007/978-3-030-96314-9_13

of the summer term was postponed due to the pandemic. The students could not return to the campus. The goal of Chinese universities was to provide flexible online education for more than 270 million students. The faculties immediately drew up plans for the implementation of online teaching, which covered the methods of teaching, managing and supervising online classes, and evaluating education [5].

There are several measures that prevent the spread of Covid-19 in schools, such as:

- Testing, quarantine, and monitoring of students and staff
- Vaccination of employees
- Ensuring physical distance, increasing hygiene standards, need to wear face masks or respirators
- Provision of resources, infrastructure for the protection of health and safety of school staff
- A detailed discussion of the measures in place
- Increase of hygiene and daily cleaning procedures
- Screening and care for sick students, teachers, and other staff and other measures [2, 3]

Social distance is one of the most effective tools for reducing the spread of coronavirus, which has caused the closure of individual establishments and institutions in most countries. Individual organizations are investigating and analyzing how the Covid-19 pandemic has affected learning curves. Learning curves are compiled by individual assessment organizations, such as the National Assessment of Educational Progress (NAEP), or international organizations such as the Program for International Student Assessment (PISA), Trends in International Mathematics and Science Study (TIMS), and Progress in International. These institutions assume three scenarios of the learning curve: a lower mean, a higher standard deviation, or a sharp increase in the low level of learning (see Fig. 1) [4].

Each scenario is caused by a different mechanism that influences students. The first scenario is the most direct transformation, which is caused by a decrease in average learning levels (blue curve). This curve represents the most likely scenario, despite the maximum efforts of educational institutions offering distance education. Students learn less despite the great efforts of teachers. The second curve represents

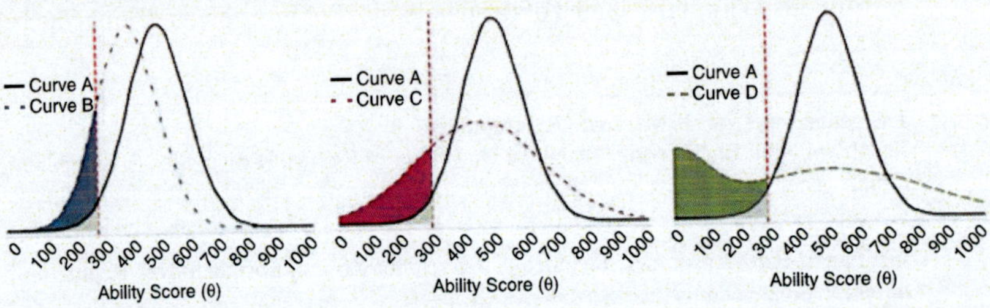

Fig. 1 Three possible scenarios of how the learning curve may evolve in the coming months [6]

a variant of the alignment or distortion of the curve due to the very unequal effects of the crisis (purple curve). Students who study regularly will progress, but stagnant students will continue to lag behind. The third curve shows the development of the learning curve due to early school leaving (green curve). The outflow of students can lead to a reduction in funding, which will lead to school closures [4].

The impacts of the Covid-19 pandemic on education and overall student life are significant. There are an increasing number of international studies examining the change in education during the global Covid-19 pandemic.

2 Innovative Technologies and the Educational Process

At present, a significant increase in digitization is across individual segments of society. Increasingly, innovative digital tools and aids have entered the individual processes. It was the pandemic that accelerated the implementation of selected innovative tools in the educational process. The ways and methods of the educational process have changed. Increasingly, m-learning and e-learning tools began to be used.

The beginning of development in the field of distance (electronic) education and digitization began with the development of the first static documents. These materials were provided to students through various tools, platforms, external memories, and the like. Subsequently, dynamic interactive solutions, platforms, and systems were developed that supported distance learning and learning processes [7].

By increasing the use of interactive solutions, information and communication technologies have been developed that have enabled the virtual expansion of education and training [8, 9].

The use of mobile platforms and smartphones for various purposes, such as e-business, process management, or distance learning through e-learning and m-learning tools, is currently increasing [10].

M-learning presents learning processes via the Internet or networks using personal mobile devices—smartphones, tablets, and laptops.

The use of mobile devices is a strong factor in improving the management of education and reducing the rate of early school leaving [11].

M-learning uses "mobile learning" for dissemination, i.e., tools such as videos, animations, tutorials, games, quizzes, interactive assessments, and the like. Elements and basic tools of m-learning include discussion forums, chats, and interactive tools for sharing information and content (information and communication tools). The main advantages of m-learning are:

- Mobility: Students can study at any place and at any time. M-learning provides instant access to relevant content. In addition, there is an increase in the possibility of interactive communication (online chat) and the possibility to participate in quizzes at your own discretion and gain access to instructional videos, which can also be watched off-line.

- Support for independence: Students learn at different rates. In face-to-face teaching, pressure is put on students to understand a particular concept or lesson within a particular lesson, which often causes frustration. Mobile learning eliminates this challenge and students learn at their own pace.
- Support for dynamism: Each student has his or her own way of learning—visual aids, audio aids, and the like. M-learning allows teachers to insert a variety of images, presentations, audio files, and videos, allowing them to adapt their interpretation style to a wide range of students. In other words, m-learning supports all types of media and creates a dynamic and engaging experience.
- Better completion rates and higher retention: Online courses are more concise and the structure of the information provided is more dynamic and transparent, which helps to better understand and store information.
- Support for teamwork: Group assignments and other advantages [12].

M-learning contains three parts:

- Learning management system (LMS) represents the functions of student management and courses, student evaluation, monitoring and their activities, etc.
- Learning content management system (LCMS) includes functions for creating content, courses, their import, export, and sharing.
- Information and communications technology (ICT) includes communication tools to support synchronization and asynchronization of teachers and students 12].

3 Methodology

3.1 Research Aim

The research was focused on the issue of teaching evaluation at the Faculty of Civil Engineering of the Technical University of Košice before and during the Covid-19 pandemic. The aim of the chapter is to compare the students' approach to study and to point out the study subjects that were most analyzed and commented on within the evaluation. Part of the research focused on the evaluation of distance learning in two periods of the pandemic (first wave—spring and summer 2020, second wave—autumn and winter 2020–2021).

3.2 Data Collection and Research Sample

As part of the research, students of the Faculty of Civil Engineering TUKE were addressed. In the period before the pandemic (academic year 2019/2020) 668 students were contacted and in the period during the pandemic (academic year

2020/2021) 723 students were contacted. Of the total number of students, the return was at the level of 39.67% (academic year 2019/2020) and 11.89% (academic year 2020/2021). The return rate of the questionnaire in the second year (during the pandemic) was relatively low, probably due to poorer awareness and reluctance of students to complete the questionnaire. The data were obtained through an online questionnaire. The research sample was contacted by email and via the MAIS information system.

3.3 Research Step and Methodology

The first step of the research was to determine the problem based on a theoretical analysis. Research questions and hypotheses were set in this area. In the introductory part of the questionnaire for students, the respondent was acquainted with the purpose of the questionnaire and the approximate time frame for completion. The questionnaire itself consisted of nine questions and was divided into four sections, concretely:

- First section: information about the respondent—study program, year, form of study
- Second section: approach to study—subjective evaluation: students' opinion
- Third section: evaluation of subjects—definition of a "significant" subject: positive or negative experience
- Fourth section: evaluation of distance teaching—form of teaching, content of teaching, and students' approach to study during individual waves of the pandemic

4 Results

4.1 Information About the Respondent

In the first part of the research, data on respondents were obtained. The research was carried out in two periods, namely in the winter semester of the academic year 2019/2020 and in the winter semester of the academic year 2020/2021; therefore the number of respondents differs. In the academic year 2019/2020, 668 students were contacted with a return of 39.7% and in the academic year 2020/2021 723 students were contacted, while the return was 11.9%. The return of the questionnaires recorded a significant decrease in interest (feedback from students) despite the repeated calls from us to students to fill in the questionnaire.

A positive trend was observed with the rate of return in the external studying, where the rate of return increased almost 18-fold, i.e., the rate of return increased

	Academic year / Parameters	2019/2020	2020/2021	Trend analysis 2019/2020 vs. 2020/2021
	Total number of students	668	**723**	increase
	Number of responses	**265**	86	decline
	Number of responses (%)	**39,7%**	11,9%	decline
	Form of study (full-time study/ part-time study)	**261** / 1	80 / **6**	decline / increase
	Form of study % (full-time study/ part-time study)	**99,6%** / 0,4 %	93% / **7%**	%
Representation	1st year - bachelor study	149 (56,4%)	23 (26,7%)	decline
	2nd year - bachelor study	34 (12,9%)	19 (22,1%)	increase
	3rd year - bachelor study	5 (1,9%)	11 (12,8%)	increase
	4th year - bachelor study	6 (2,3%)	4 (4,7%)	increase
	1st year - engineering study	42 (15,9%)	12 (13,9%)	decline
	2nd year - engineering study	28 (10,6%)	17 (19,8%)	increase
	3rd year - engineering study	0 (0%)	0 (0%)	without a change

Legend: increase decline without a change

Fig. 2 Description of the research sample and analysis of trends

from 0.4 to 7%. In both periods, all years of full-time study were included in the survey. A detailed analysis of the research sample is shown in Fig. 2.

In terms of the representation of study programs, in the monitored periods 2019/2020 and 2020/2021, a percentage increase in return (filling in the questionnaire) was recorded in the study programs civil engineering and architecture (Bac. study), civil engineering (Eng. study), engineering structures and traffic structures (Bac. study), load-bearing structures of building (Eng. study), technology and management in construction (Bac. study and Eng. study), and building with environmental purpose (Bac. study and Eng. study). Detailed analysis of trends—return within study programs—is shown in Fig. 3.

4.2 *Approach to Study and Analysis of Subjects: Results*

Another part of the research was focused on the analysis of students' approach to study. Students answered the simple question "How do you evaluate your approach to study?" The student's attitude to the study is influenced by many aspects. One of the important factors in building motivation and desire to learn and work is the

Study program	Number of responses (%) 2019/2020	Number of responses (%) 2020/2021	Trend analysis 2019/2020 vs. 2020/2021
civil engineering and architecture (Bc. study)	45,1%	45,4%	↗
civil engineering (Eng. study)	9,5%	13,9%	↗
technical equipment of buildings (Eng. study)	6,4%	0,0%	↙
engineering structures and traffic structures (Bac. study)	3,4%	3,5%	↗
load-bearing structures and transport structures (Eng. study)	1,1%	2,3%	↗
load-bearing structures of buildings (Eng. study)	1,5%	1,2%	↙
technology and management in construction (Bac. and Eng. study)	29,2%	31,4%	↗
realization of transport constructions (Bac. study)	1,9%	0,0%	↙
buildings with environmental purpose (Bac. and Eng. study)	1,9%	2,3%	↗
buildings for sustainable water management in the country (Bac. and Eng. study)	0,0%	0,0%	—

Legend: increase decline without a change

Fig. 3 Return of the questionnaire from the point of view of study programs—trend analysis

approach of teachers to teaching and students. It is the friendly approach and a good teaching atmosphere that is a positive factor for students, which influences their overall approach to study. Due to the fact of a change in the form of teaching, from full-time to distance form (effects of the pandemic), various innovative technologies, i.e., m-learning tools, were implemented more quickly into the educational process. Various information and communication tools have been used to a greater extent, such as Cisco Webex and the TUKE Moodle tool, which is a learning management

Approach to study	Number of responses				Trend analysis 2019/2020 vs. 2020/2021
	2019/2020		2020/2021		
	number	%	number	%	
1 - very lax approach	12	4,5%	3	3,5%	↙
2 - lax approach	10	3,8%	0	0,0%	↙
3 - neutral approach	63	23,8%	12	14,0%	↙
4 - responsibly approach	92	**34,7%**	31	**36,0%**	↗
5 - very responsibly approach	88	**33,2%**	40	**46,5%**	↗

Legend: ↗ increase ↙ decline

Fig. 4 Approach to study: subjective evaluation of students

system (LMS) used to support and improve the quality of teaching. In this system, students are provided with study materials, audio recordings, and video recordings, and within the system, it is also possible to test students and online communication. Cisco Webex Information and Communication Tool is an easy-to-use and secure application for calling, messaging, meeting meetings, and working. The use of this platform increases the connection with students through videoconferencing, and increases cooperation and interactivity between students and teachers. The utilization rate of these tools increased by almost 70% compared to the pre-pandemic period (academic year 2019/2020) and during the pandemic (academic year 2020/2021). Based on the above facts, the students' approach to study has also changed. The survey showed the expected trend that as digitalization and use of innovative technologies and platforms increase, satisfaction will also increase and students' access to study will improve. This trend has been confirmed and the results are shown in Fig. 4.

Part of the research was also focused on the evaluation of individual subjects of study. In both cases, the most evaluated and analyzed subjects were focused on mathematical disciplines such as mathematics 1, mathematics 2, numerical mathematics, and mathematical statistics and descriptive geometry (period 2019/2020—42.64% of respondents, period 2020/2021—38.37% of respondents).

Of the professional subjects, subjects focused on architectural design of structures were most analyzed in both periods, for example building studio, building structures, typology, and the like (period 2019/2020—7.55% of respondents, period 2020/2021—17.44% of respondents). In the period of the winter semester 2019/2020, the subject technical equipment of buildings (3.1% of respondents) was also significantly evaluated, and in the period of the winter semester 2020/2021 the subject technology of construction processes (9.3% of respondents). Students in the mentioned subjects evaluated positively especially the implementation of digital tools into individual processes.

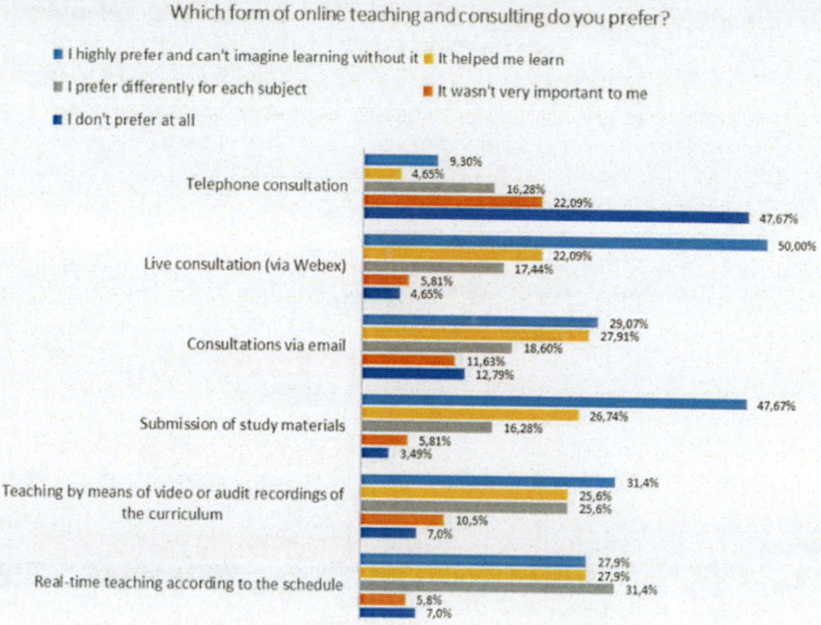

Fig. 5 Results: "Which form of online teaching and consulting do you prefer?"

4.3 *Evaluation of Distance Learning: Results*

The last part of the research was focused on the evaluation of distance education. Students defined their preferences in the areas of:

- Real-time teaching according to the schedule
- Teaching by means of video or audio recordings of the curriculum
- Submission of study materials
- Consultations via email
- Live consultation (via Webex)
- Telephone consultation

The results are shown in Fig. 5. Students did not prefer telephone consultations at all, and the most preferred elements of distance learning were teaching through video or audio recordings of the curriculum, sending study materials, and live consultations through Cisco Webex tools. With regard to real-time teaching, the preferences for this method of teaching varied according to the subject.

Another part of the research was focused on the evaluation of distance education during the first and second waves of the Covid-19 pandemic, i.e., a comparison of the summer semester of the academic year 2019/2020 with the winter semester of the academic year 2020/2021. Students commented on the form of teaching (use

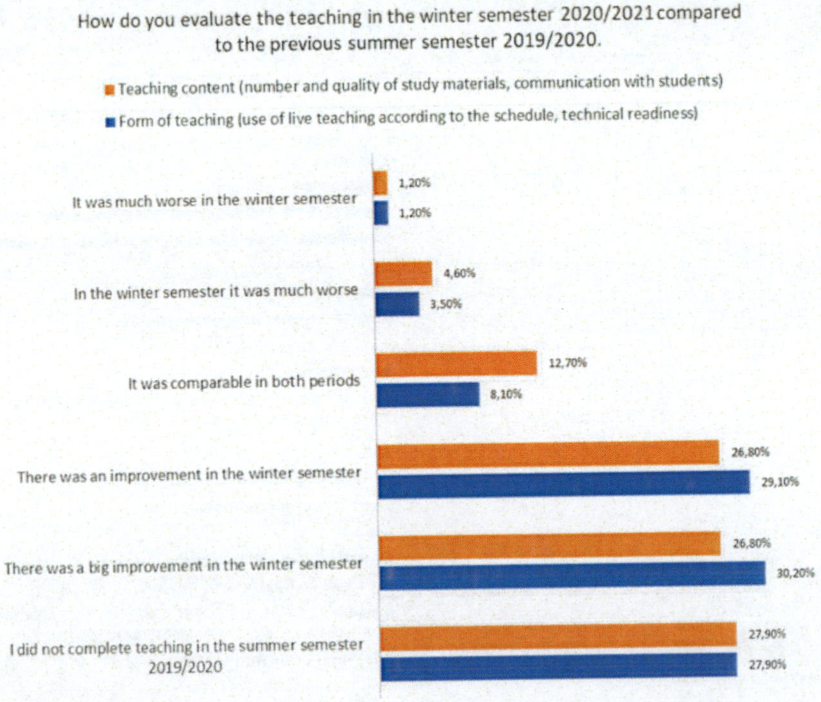

Fig. 6 Results: "How do you evaluate the teaching in the winter semester 2020/2021 compared to the previous summer semester 2019/2020?"

of live teaching according to the schedule, technical readiness) and the content of teaching (number and quality of study materials, communication with students). From the above points of view, according to the students, there was an improvement during the second wave of the pandemic; detailed results are shown in Fig. 6.

The last part of the research analyzed students' approach to teaching during the Covid-19 pandemic, namely the comparison of the first wave of the pandemic (summer semester 2019/2020) with the second wave of the pandemic (winter semester 2020/2021).

The obtained data confirm the statement that with the increase in the use of digital technologies and the implementation of innovative elements, the level of student satisfaction increases. When comparing the different periods, the students stated that either their approach did not change at all (25.6%) or there was an improvement (see Fig. 7).

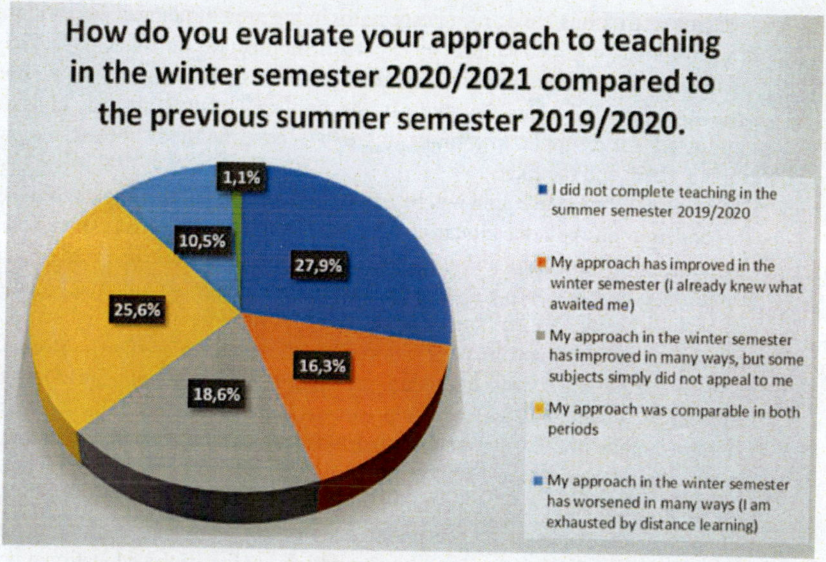

Fig. 7 Results: "How do you evaluate your approach to teaching in the winter semester 2020/2021 compared to the previous summer semester 2019/2020?"

5 Conclusions

The Covid-19 pandemic has affected all aspects of life and changed the way many institutions and organizations operate. The effects of the pandemic have severely affected the educational process. Individual educational institutions—preschools, primary schools, high schools, and universities—had to be closed and switch from full-time to distance learning. The aim of the research was to analyze the form and content of teaching before and during the Covid-19 pandemic at the Faculty of Civil Engineering in Košice. As a result of the pandemic, the implementation and use of selected innovative technologies and processes in the educational process have increased rapidly. Various information and communication tools and the learning management system were used to a greater extent. Students were provided with study materials, audio recordings, and video recordings, and within the abovementioned tools, credit papers, exams, and final defenses were also carried out. The obtained data confirm the assumption that with the use of digital technologies and innovative elements, the level of student satisfaction increases. Satisfaction rates increased significantly during the second wave of the pandemic. This was helped by the fact that students already knew what to expect and that the form and content of the study improved. Teachers used more innovative tools and digital technologies, which increased the flexibility and attractiveness of the subject—adaptation of content to the listener, better presentation, use of digital tools such as virtual reality, building information modelling tools, and other innovative technologies and tools.

Great restrictions have arisen, especially in the area of connection with practice. Many teachers provided lectures by experts from practice, but students lacked direct interaction on construction sites, in construction companies, and in laboratories. From this point of view, the study in the field of construction was considerably eliminated, for example, in contrast to the IT sector, the medical field, the pharmaceutical field, and the like.

The aim of future research will be a thorough analysis of the impact of the Covid-19 pandemic on the educational process. Given the possible risks of reincrease in the number of infections and the subsequent introduction of restrictive measures, the aim of future research will be to create the so-called crisis automat for education which will be the plan of steps for the educational phases with regard to region differences as in official COVID automat in Slovakia—five-color system. On the basis of the obtained data, individual procedure steps, methods, and tools will be determined, which will directly help to increase the efficiency, productivity, and quality of education during the crisis period. At first, the educational automat will be an internal tool of the Faculty of Civil Engineering in Košice. The individual steps, methods, and tools of teaching will be individually configured for individual subjects, with emphasis on the preferences of students and teachers in order to ensure the smooth running of teaching and high quality of the educational process.

Acknowledgements The chapter presents partial research results of the project "Use of building information model in planning the life cycle of buildings" and this work was supported by the Slovak Research and Development Agency under the contract no. APVV-17-0549.

References

1. UNDP, Covid-19 pandemic—Humanity needs leadership and solidarity do defeat the coronavirus (2021), https://www.undp.org/content/undp/en/home/coronavirus.html. Accessed 26 Apr 2021
2. World Health Organization, Coronavirus disease (COVID-19): Schools (2020), https://www.who.int/emergencies/diseases/novel-coronavirus-2019/question-and-answers-hub/q-a-detail/coronavirus-disease-covid-19-schools#. Accessed 22 Apr 2021
3. A.R. Grahl, Educating in a pandemic and beyond, in *Rotary* (2020), https://www.rotary.org/en/educating-in-a-pandemic-and-beyond?gclid=CjwKCAiAouD_BRBIEiwALhJH6IYnMGF7zkQmpypBaRK3Riy4tnYn6Q03XCiKJHbG9ym1r7IrzMWb4RoCP-oQAvD_BwE. Accessed 22 Apr 2021
4. S.A. Iqbal, J. P. Azevedo, K. Geven, A. Hasan, H.A. Patrinos, We should avoid flattening the curve in education—Possible scenarios for learning loss during the school lockdowns (2020), https://blogs.worldbank.org/education/we-should-avoid-flattening-curve-education-possible-scenarios-learning-loss-during-school. Accessed 22 Apr 2021
5. X. Zhu, J. Liu, Education in and After Covid-19: Immediate Responses and Long-Term Visions (Nature Public Health Emergency Collection, 2020). https://doi.org/10.1007/s42438-020-00126-3. https://www.ncbi.nlm.nih.gov/pmc/articles/PMC7183744/. Accessed 27 July 2021
6. Blogs.worldbank.org: Three possible scenarios of how the learning curve may evolve in the coming months (2020), https://blogs.worldbank.org/sites/default/files/styles/full_width/public/blogs-images/2020-04/blog_graph_2.png?itok=EotqJkal. Accessed 22 Apr 2021

7. E.A. Cudney, S.L. Murray, B. Groner, K.M. Kaczmarek, B., Wilt, K. Blaney, J. Phelps, Using the Voice of the Student to Evaluate Learning Management Systems. American Society for Engineering Education (2017), https://scholarsmine.mst.edu/cgi/viewcontent.cgi?article=1459&context=engman_syseng_facwork. Accessed 27 July 2021
8. W. Guo, *Improving Engineering Education Using Augmented Reality Environment*, in International Conference on Learning and Collaboration Technologies (Springer, Cham, Switzerland, 2018), pp. 233–242
9. Y. Aliev, V. Kozov, G. Ivanov, A. Ivanov, *3D Augmented Reality Software Solution for Mechanical Engineering Education*, in Proceedings of the 18th International Conference on Computer Systems and Technologies, Ruse, Bulgaria, 23–24 June 2017 (ACM, New York, NY, 2017), pp. 318–325
10. M. Jou, Y.T. Lin, H.C. Tsai, Mobile APP for motivation to learning: an engineering case. Interact. Learn. Environ. **24**, 2048–2057 (2016)
11. O. Moscoso-Zea, P. Saa, S. Luján-Mora, Evaluation of algorithms to predict graduation rate in higher education institutions by applying educational data mining. Australas. J. Eng. Educ. **24**(1), 4–13 (2019). https://doi.org/10.1080/22054952.2019.1601063
12. H. Singh, Why Mobile Learning Is Now More Than Just A Nice-To-Have, 10 (2020). eLearningindustry.com. https://elearningindustry.com/why-mobile-learning-is-more-than-just-nice-to-have. Accessed 26 Apr 2021

Application of the Vernier Measuring System as a Smart Element in Physics Education Process

Patrik Kaščák, Martin Miškiv-Pavlík, Denisa Olekšáková, and Jozef Jurko

1 Introduction

These days there is much lay and professional discussion about the best way and level of teaching physics in schools, including the extent of the curriculum for students. It is common knowledge that physics is not one of the most popular subjects in elementary and high schools. The main reason for this is often the problem that students do not see its practical significance and application. It should not be forgotten that organizing an experimental lesson for students requires proper training by the teacher. Not to mention the time intensity, inadequate material equipment of the school, insufficient number of tools, and poor working environment. In the world, but also in Slovakia, the trend is towards innovation and modernization of the educational process and the introduction of experimental and innovative technologies into the educational process. Therefore, the issue of building or equipping of excellent physical laboratory is currently a topical problem.

The main thesis of this chapter is to apply Vernier's experimental measurement systems in the physics teaching process at a university with a technical emphasis and to demonstrate the positive impact of this method on physics education and laboratory modernization on student's assessment results after their application through statistical analysis. Based on the students' responses to each question in the questionnaire, the teacher can then identify the area with which the students have the greatest problem and explain the problem using individual Vernier measurement systems.

P. Kaščák (✉) · M. Miškiv-Pavlík · D. Olekšáková · J. Jurko
Faculty of Manufacturing Technologies with the Seat in Prešov, Technical University of Košice, Prešov, Slovakia
e-mail: patrik.kascak@tuke.sk

L. Knapčíková, D. Peraković (eds.), *6th EAI International Conference on Management of Manufacturing Systems*, EAI/Springer Innovations in Communication and Computing, https://doi.org/10.1007/978-3-030-96314-9_14

1.1 Experimental Education of Physics in Practice

At present, there are many publications on applied experiments, tutorials, and new methods, but in the field of application, no special attention is currently paid to experimental equipment in the physical laboratory. Kanim and Cid in their study conducted a large experiment on physics knowledge of students in the United States, dividing and testing students according to whether they studied physics in technical or humanities schools. The results of the tests showed that technical high schools, where some of the experimental teaching methods were used, had better test scores than students from humanities schools, where physics instruction was only at the theoretical level [1].

Treagust and Duit in their thesis deal with the description of some teaching aids for students, demonstration systems, and their impact on the teaching process of physics in high schools [2]. Suhendi and Ramdhani in their publication also deal with issues of organization of the laboratory and its equipment in experiments in physics in all types of schools according to the difficulty of implementation [3]. Hodosyová and Útla in their research state that the positive benefit of experimental learning is that students become more active in a new or well-equipped laboratory, as this environment stimulates their curiosity and interest, and the implementation of experiments [4] (Fig. 1).

Darmaji and Kurniawan in their study suggested a change in the cooperation system of school education with real life and the idea of widening the space for individuality with the principle of not learning everything but focusing on specific cases of applied physics in the subject under study [6]. Etkina and Gregorcic in their research point out that by proper use of experimental tasks we support creative thinking of students and force them to solve problems, find answers, ask questions, and justify their opinions, which are important aspects for young people to assert

Fig. 1 Laboratory of applied physics with Vernier's measuring system [5]

themselves in this time [7]. With an interesting thesis came Maulidah and Prima who created a virtual model in the process of propagation of sound waves [8]. Cebrian et al. in their research describe the benefits of a virtual lab for students with the ability to self-select [9]. In their publication, Liang et al. made a proposal in which they taught physics to students via smartphones, thanks to a software they created themselves, where the students could see real test results on their phones [10].

1.2 Vernier Measuring System

Nowadays engineering practice requires highly skilled professionals who have a comprehensive understanding of the state of modern engineering and can work with the latest technologies. That is the reason why a modern applied physics laboratory has been set up at the laboratory of applied physics in this chapter, which uses the latest information and education technologies as well as vernier measurement systems [11].

The education process of subjects is innovated by the mentioned Vernier system, which is a complete solution for school laboratories and allows practical teaching with the latest computer and mobile communication devices. This system is characterized by very simple and intuitive operation, a wide range of different measurement sensors, and reliable software solutions. The results of laboratory measurements can be evaluated not only on computers, but also on tablets or smartphones [11] (Fig. 2).

The measuring systems also include the GDX system for investigating the damping force and its relationship to angular velocity, weight, and turning radius.

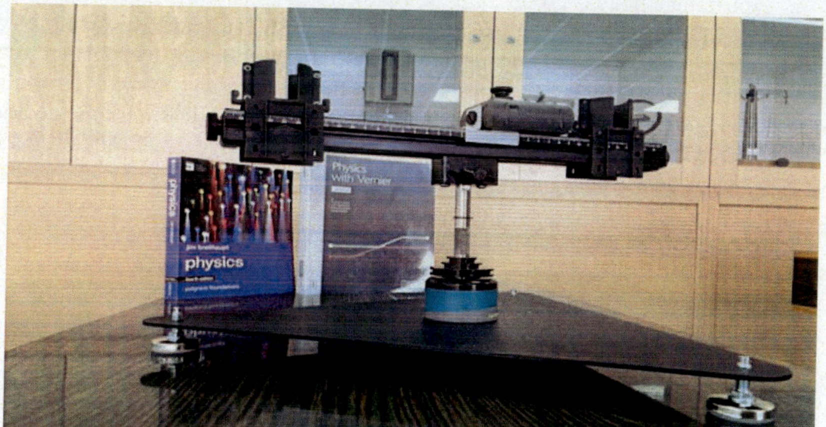

Fig. 2 Image of measuring GDX system for centripetal force [5]

Fig. 3 Movable trolley on GDX system frame [5]

The swivel arm with two movable trolleys and the cable mechanism is mounted on a sturdy frame. Students can use a single system to make various measurements of rotational dynamics, such as measuring angular velocity $\omega = d_{\varphi}/d_t$, centrifugal force $F_d = (m.v^2)/d$, or acceleration $a_d = v^2/d$ [11].

We can also study rotational dynamics by calculating the forces necessary to keep matter in circular motion $F = m.\ a$ or Newton's laws which involve torque and rotation motion [11] (Fig. 3).

The second of the applied measurement systems is a dynamic wheelchair system with wireless connection to study the motion, flexible and inflexible collisions of the carriages, and laws of conservation of momentum and energy [11].

DTS-GDX wheelchairs have built-in position sensors that work on the principle of distance measurement by using a special rotating part that rolls on the rail, a force sensor, and a three-axis accelerometer and a spring piston when reflecting. DTS-GDX wheelchairs can be used on the rolling track, but also on any other flat and smooth surface (table, board) [11].

The system consists of a 1.2 m track, a trolley with magnetic buffers and Velcro buffers, four additional trolley weights, mounting accessories for motion sensors, photoelectric sensors, force sensor and acceleration sensors, reflective surface for the motion sensor, a low-friction idler pulley, and an idler pulley holder [11] (Fig. 4).

A typical example of the application of this measurement system of dynamic wheelchairs can be in an experiment with an inclined plane, where we can monitor

Fig. 4 DTS measuring system [12]

Fig. 5 Application of the DTS measuring system in the laboratory [5]

the basic motion of the carriage, or Galileo's experiment of determining *g* with a body on an inclined plane [11].

Another possibility is to study Newton's second law using a force sensor on a trolley. We can record both the acting force and the acceleration, which are directly proportional. Using a force sensor and a set of buffers and reflections, it is possible to study the relationship between momentum and impulse. The momentum is determined by integrating the force-time curve [11] (Fig. 5).

2 Research Methodology

The experimental part of this chapter consists of the study and statistical processing of the results of students in the first year of bachelor study at the beginning of the semester physics course with the Vernier measurement system and after its application in the educational process. The statistical processing itself consisted of the processing of two questionnaires. The entry questionnaire was completed at the beginning of the semester when the students had no experience in using the Vernier measurement system and only had knowledge from high school. In comparison to the entry questionnaire, a second output questionnaire was also completed after which we evaluated the students' score results from questions based on Grubbs's and Dixon's statistical processing [13].

2.1 *Grubbs's Test*

Sometimes there are values among the measured values that differ greatly from others. We call such values extreme values (outliers). Extreme values distort the calculated properties of random selection, which can lead to an incorrect conclusion when analyzing a given quantity or identifying a particular value, such as determining the difficulty of a test question based on the number of correct answers. The Grubbs's test is one of the most commonly used tests for extreme values [13].

This statistical method assumes that the population distribution is satisfied with a small random selection range ($n \leq 30$). The values of the random selection are to be arranged in order of magnitude in a nondecreasing sequence (coefficient series):

$$x(1) \leq x(2) \leq x\,(n-1) \leq x(n)\ [13]\,.$$

The Grubbs's test itself consists of two forms, depending on whether we are testing the minimum or maximum extreme value. When testing the extreme value of the minimum value, the test criterion is:

$$T(1) = \frac{\overline{x} - x(1)}{s}\sqrt{\frac{n}{n-1}}\ [13]\,,$$

or

$$T(n) = \frac{X_{(n)} - \overline{x}}{s}\sqrt{\frac{n}{n-1}}\ [13]\,,$$

where:

$\overline{x}$ is the arithmetic mean of the sample
s is the standard deviation of the sample
n is the range of random selection

We reject the null hypothesis H0 at the significance level α, if

$$T(1) > T_\alpha(n), T(n) > T_\alpha(n)\,[13],$$

where T_α is the pane critical value from the Grubbs's test schedule [13].

2.2 Dixon's Q Test

Like the Grubbs's test, the Dixon's Q test has two forms, depending on whether we are testing the extremity of the minimum or maximum value. In this case, it is not necessary to calculate the arithmetic mean of the sample and the selected standard deviation of random selection. To calculate this method use the formula:

$$Q(1) = \frac{X_{(2)-X_{(1)}}}{X_{(n)} - X_{(1)}}\,[13],$$

or

$$Q(n) = \frac{X_{(n)-X_{(n-1)}}}{X_{(n)} - X_{(1)}}\,[13],$$

depending on whether we are testing the remoteness of the smallest value or the highest value; we reject the *H0* null hypothesis at α if

$$Q(1) > Q_\alpha(n), Q(n) > Q_\alpha(n)\,[13],$$

where Q_α is the prescribed critical value from the Dixon's test table for $\alpha = 0.05$ and $\alpha = 0.01$ [13].

3 Results of Experiment

As mentioned in the Chapter "Challenges and Benefits for Detecting Soon-to-Fail Drives in Industry 4.0", the statistical processing of student results involved two types of questionnaires. The entrance questionnaire, which students took at the beginning of the semester, and the outcome questionnaire, which was administered at the end of the semester.

The entrance questionnaire consisted of six questions from different areas of physics. A total of 87 students, both male and female, participated. These students were divided according to the type of high school they graduated, namely gymnasium, secondary school, or other education institution. The students were also

Table 1 Sample of 10 students from the entrance questionnaire

Entrance questionnaire						
Student	Question 1	Question 2	Question 3	Question 4	Question 5	Question 6
1.	1	1	0	1	0	0
2.	1	1	1	1	1	1
3.	1	1	1	1	1	1
4.	1	1	1	1	1	1
5.	1	1	1	1	1	1
6.	1	0	1	1	1	1
7.	1	1	1	1	1	1
8.	1	1	1	1	1	1
9.	1	1	1	1	1	1
10.	1	0	1	1	1	1

divided according to the number of years completed in high school in physics. All test results were processed in Excel for further statistical processing.

In Table 1 we see a sample of ten students and their success rate from the entrance test, where we marked the subjects with the numbers 1–87 and for each question, they could get 1 point for the correct answer and 0 points for the wrong one. In the same way, the output questionnaire was scored on the same sample of students. From these responses, we then used the first step to create descriptive statistics to find out how successful students were in each question after graduating from different types of secondary schools. Subsequently, we also divided the students according to the number of years of physics completed in secondary school, which provided us with additional data for statistical processing according to Grubbs's and Dixon's test procedures.

3.1 Descriptive Statistics

When we find the values of a certain quantitative character on the samples of a statistical file, we obtain a set of numerical data. If the range of the statistical file is n, we get a sequence of values. It is called the primary table. The value x_1 is the minimum value of the character and x_n is the maximum value of the character. The difference $R = X_{HAX} - X_{HIH}$ between the maximum and minimum values of a character is the so-called range (or coefficient width) of the character. At the same time, we can tell from the ordered table whether individual character values occur more than once. If this is the case, then we can create a so-called tab to divide the numbers. The first column contains different character values, and the second column their (absolute) number. The symbol f_i indicates the frequency of occurrence of $x_i (i = 1,2\ldots,k)$. The following applies: $f_1 + f_2 + \ldots + f_k = n$. We can add more columns to a table of number distributions, a column of relative numbers, a column of cumulative numbers, and a column of cumulative relative numbers.

Table 2 Distribution of the number of measured values from entrance questionnaire

Number of points x	Absolute abundance f_i	Cumulated abundance	Relative abundance f_i/n	Relative abundance in %	Cumulative relative abundance
1	0	0	0	0	0
2	0	0	0	0	0
3	0	0	0	0	0
4	10	10	0.12	12	0.12
5	49	59	0.56	56	0.68
6	28	87	0.32	32	1
Sum.	87	–	1	100	–

Table 3 Distribution of the number of measured values of the output questionnaire

Number of points x	Absolute abundance f_i	Cumulated abundance	Relative abundance f_i/n	Relative abundance in %	Cumulative relative abundance
1	1	1	0.015	1,5	0.015
2	1	2	0.015	1,5	0.03
3	2	4	0.02	2	0.05
4	12	14	0.14	14	0.19
5	22	36	0.25	25	0.44
6	49	87	0.56	56	1
Sum.	87	–	1	100	–

As a first step in the statistical analysis, we started with a comparison of the individual test results. The results of the input and output tests were divided in a simple breakdown according to the number of correct answers of each student, from 1 to 6, representing the value of x.

Analysis of Table 2 showed that the highest number of responses ranged from 5 to 49 students, representing 56% of the total number of responses. Overall, 68% of the students had 5 correct answers and only 32% of the students scored full marks. Compared to the baseline test from Table 3, there was an improvement in students' overall scores when the highest number of correct responses was 6, with 49 students representing 56% of the total responses, which is a 24% improvement compared to the entry questionnaire.

The percentage of students who gave up to 5 correct answers dropped to 44%, which is 12% less than the entrance questionnaire. One of the negative phenomena that occurred in the output questionnaire was also the occurrence of lower interval response categories, such as 1, 2, or 3 responses, which did not occur in the entrance test, but their number has very little impact on the total number of responses.

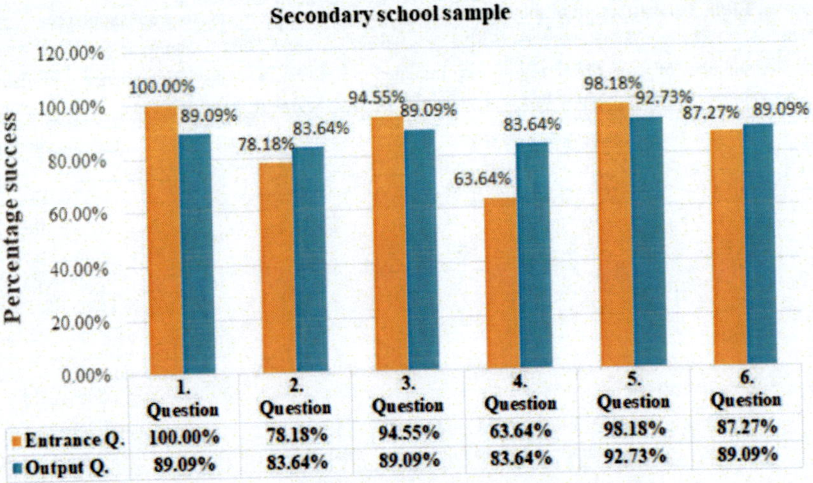

Fig. 6 Comparison of vocational school sample entry and output questionnaire results

3.2 *Vocational School Sample*

We then divided the respondents according to the secondary schools they had completed. The first group was the graduates of specialized secondary schools. This criterion included 55 respondents who were not yet subdivided according to the number of years of physics completed.

It can be seen from Fig. 6 that the biggest problem for the students in the entrance questionnaire was question 5, which only 63% of the respondents answered correctly and they performed best in question 1, where they achieved a success rate of 100%. The overall pass rate of the students was 86.9%.

Compared to the baseline questionnaire, there was an overall improvement in student scores when using the Vernier educational measurement system.

The distance learning factor due to the COVID-19 pandemic must also be factored into the questionnaire results, which was reflected in some of the poorer responses on the output questionnaire, such as question 1. The overall student success rate was 87.9%, which was 1% higher than the entry questionnaire.

3.3 *Gymnasium Sample*

The second group of respondents were students who graduated from gymnasium type of secondary school. The number of respondents this time was 13. In the case of the entrance questionnaire of students who graduated from gymnasium secondary,

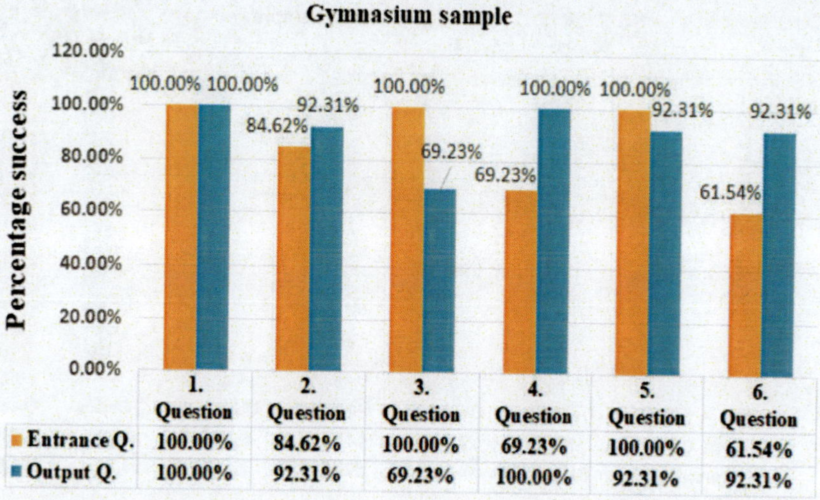

Fig. 7 Comparison of the results of the entrance and output questionnaire of the gymnasium sample

we can see from Fig. 7 that the lowest percentage of responses was for questions 6 and 4, only 61% and 69%.

Overall, the average percentage pass rate of students from the entrance questionnaire was 85.9%. Comparing the results of the entrance and output questionnaire of the gymnasium students' sample, we can see an improvement in questions 2, 4, and 6 and a deterioration in 3 and 5. The overall success rate was at an average level of 91%, which corresponds to an improvement of 5%.

By comparing the results of the entry and output questionnaire of a sample of gymnasium students, we can see an improvement in questions 2, 4, and 6 and a deterioration in 3 and 5. The overall success rate was at an average level of 91%, representing an improvement of 5%.

3.4 *Other School Sample*

As a final group of respondents, we have students who graduated from a high school other than gymnasium or vocational, such as art schools. In total, there were 19 students in this sample.

From the results in Fig. 8, we can evaluate that this sample showed the lowest or worse results when in the entrance questionnaire students scored an average percentage of 87.7% and in the output questionnaire 85.9%. The biggest problem was again made for students' question 4, and the highest number of correct answers was in question 1 (Fig. 9).

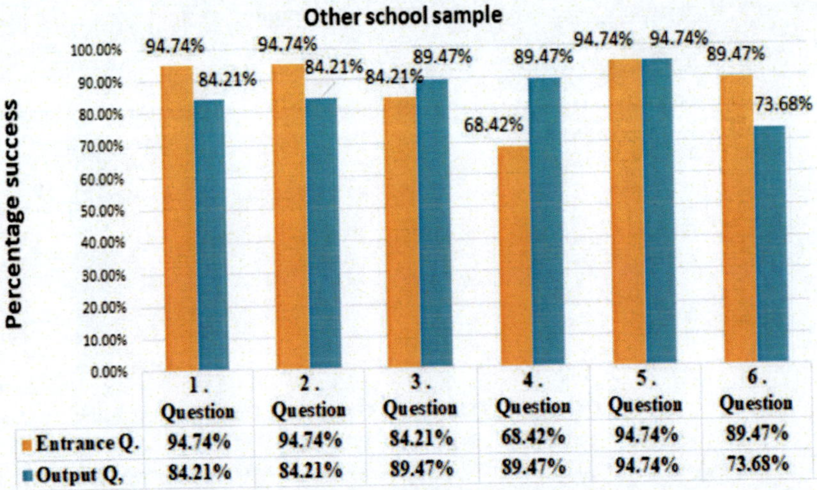

Fig. 8 Comparison of the results of the entry and output questionnaire for other school sample

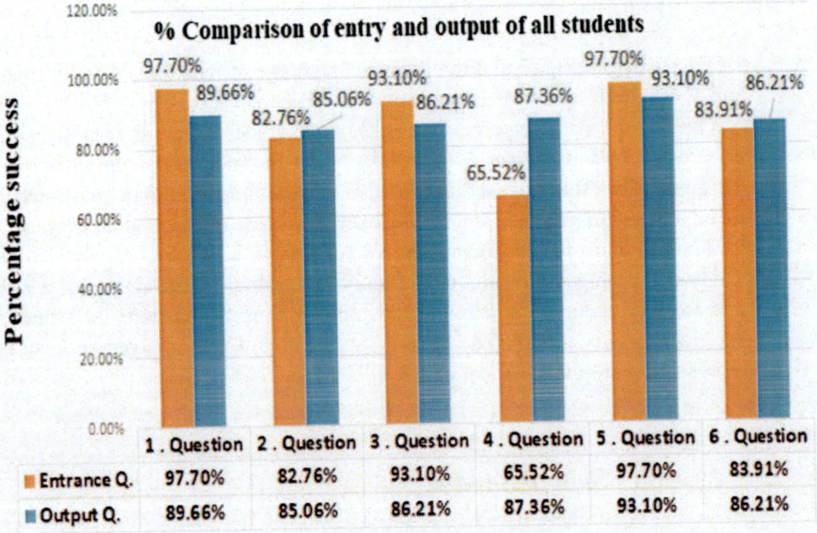

Fig. 9 Comparison of entry and output questionnaire of all students

In the overall comparison of the results from the entrance and output questionnaires, we can conclude that there was an improvement in the results, which is confirmed by the average value of the percentage of the questionnaire results, where the entrance questionnaire was 86.8% and the output questionnaire showed an improvement to 87.9%. The highest improvement occurred in question 4 when there was an improvement from 65.2 to 87.3%.

3.5 *Distribution of Results by Years of Physics in High School*

After dividing the respondents by completed years of physics in high school, we can see that there was an improvement in scores in almost every category. For students who had only 1 year of physics, mostly students from other schools, the improvement in scores ranged from 87 to 89%. In the 2-year category, the improvement was from 84.7 to 85.4%, mostly students from vocational high school.

The third category was 3-year physics. In this single category the results deteriorated from 83.3% to 80.9% with the largest number of students being gymnasts. The final category was 3+ years' completed physics where the improvement was from 88.2 to 90.2%. The composition of the students was mainly from grammar school and the minority proportion from vocational school (Figs. 10, 11, 12, and 13).

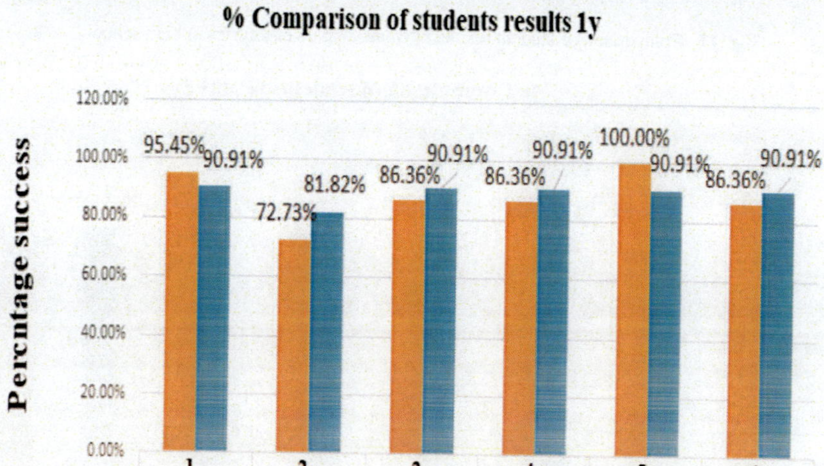

	1. Question	2. Question	3. Question	4. Question	5. Question	6. Question
Entrance Q.	95.45%	72.73%	86.36%	86.36%	100.00%	86.36%
Output Q.	90.91%	81.82%	90.91%	90.91%	90.91%	90.91%

Fig. 10 Comparison of students' results by year: 1 year of physics in high school

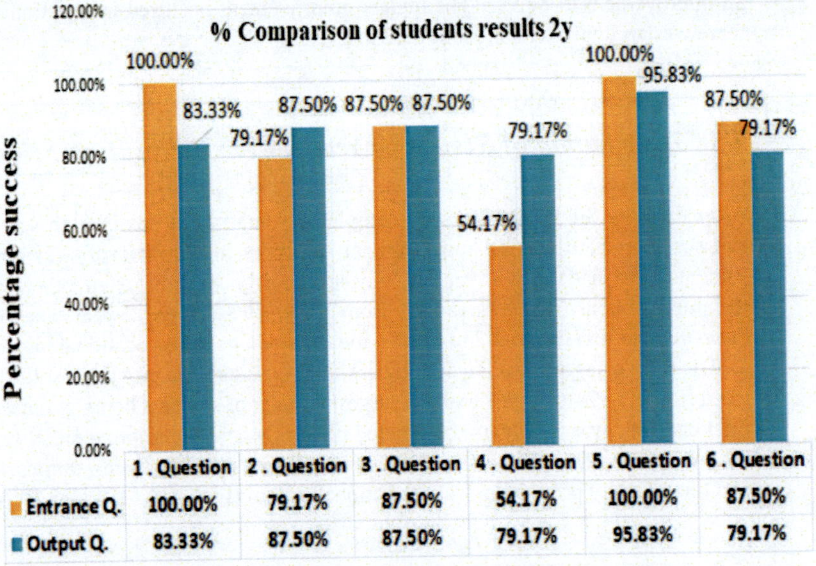

Fig. 11 Comparison of student results by year: 2 years of physics in high school

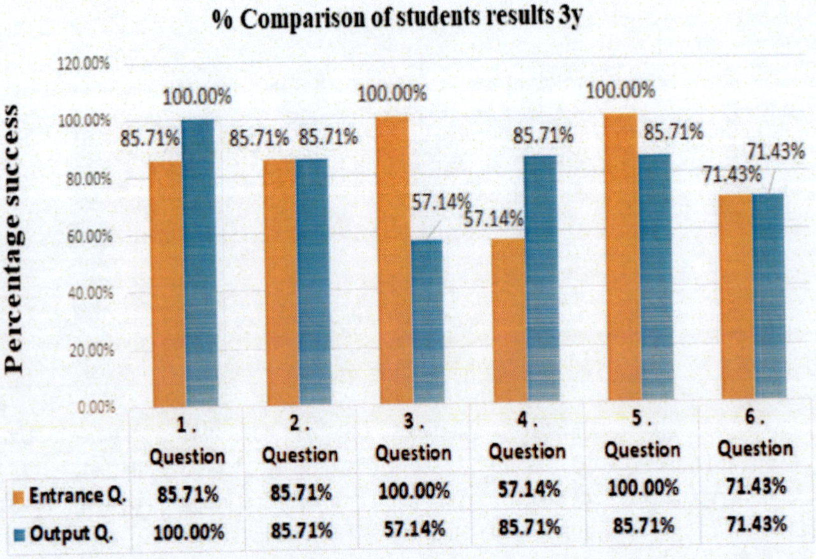

Fig. 12 Comparison of student results by year: 3 years of physics in high school

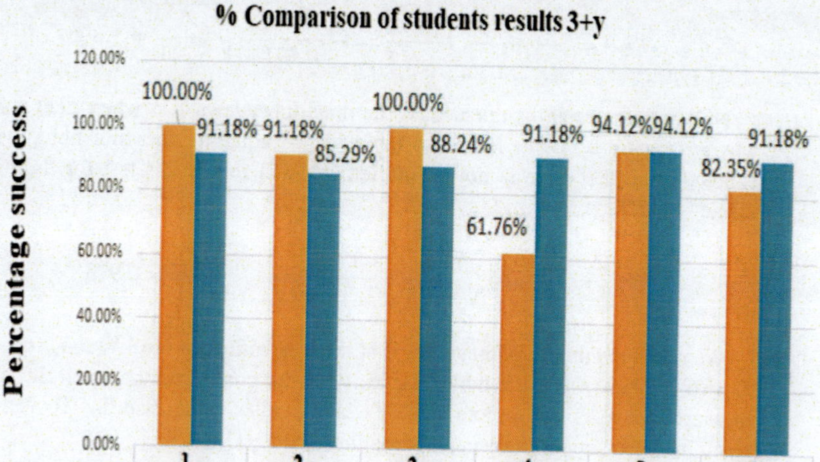

	1. Question	2. Question	3. Question	4. Question	5. Question	6. Question
Entrance Q.	100.00%	91.18%	100.00%	61.76%	94.12%	82.35%
Output Q.	91.18%	85.29%	88.24%	91.18%	94.12%	91.18%

Fig. 13 Comparison of student results by year: 3+ years of physics in high school

3.6 *Grubbs's and Dixon's Q Test of Detecting Outlier*

By using these methods, we can define whether the questionnaire results contain a value that proves to be an anomaly compared to other values. Through such analysis, more challenging questions can be defined based on the number of correct answers in the questionnaire.

As a first step, in both questionnaires we rank the sample by size into a nondecreasing sequence $x_{(2)} \leq x_{(2)} \leq \ldots \leq x_{(n)}$. The measured values for the entry questionnaire were 57, 72, 73, 81, and 85 and for the output 74, 75, 76, 78, and 81 correct responses.

The first questionnaire is to determine if the values 57 and 85 are outliers. For the Grubbs's test, the arithmetic mean of the sample is $\overline{x}$. The first questionnaire is to determine if the values 57 and 85 are outliers. For the Grubbs's test the arithmetic mean of the sample is

$$n = 6, \overline{x} = 75.5, s = 10.691.$$

We will test the smallest value first, so we will use the test characteristic at $\alpha = 0.05$, whose value is

$$T(1) = \frac{\overline{x} - x(1)}{s} \cdot \sqrt{\frac{n}{n-1}} = \frac{75.5 - 57}{10.691} \cdot \sqrt{\frac{6}{6-1}} = 1.896.$$

Null hypothesis H_0 is rejected, if the materiality level is α when $T(1) > T_\alpha(n)$, where $T_{0.05}(6) = 1.996$ from the table. Since equality does not hold, we can conclude that the value is not an outlier. We will repeat this test for the highest value:

$$T_n = \frac{x_n - \overline{x}}{s} \cdot \sqrt{\frac{n}{n-1}} = \frac{85 - 75.5}{10,691} \cdot \sqrt{\frac{6}{6-1}} = 0.973.$$

We reject the null hypothesis H_0 if at the α significance level $T_n > T_\alpha(n)$, where $T_{0.05}(6) = 1996$ from the table. Again, we were confirmed that even the highest measured value is not remote since T_n is a smaller value than T_α. To verify the result, we also support it by calculating the Dixon's Q test:

$$Q(1) = \frac{X_{(2)} - X_{(1)}}{X_{(n)} - X_{(1)}} = \frac{72 - 57}{85 - 57} = 0.534.$$

We reject the null hypothesis H_0 if at significance level α, if $Q(1) > Q_\alpha(n)$, where $Q_{0,05}(6) = 0.560$ from the table. Again, this equality does not hold, and so the Dixon's Q test confirms that the value is not removed (Figs. 14 and 15):

$$n = 6, \overline{x} = 76.5, s = 2.588.$$

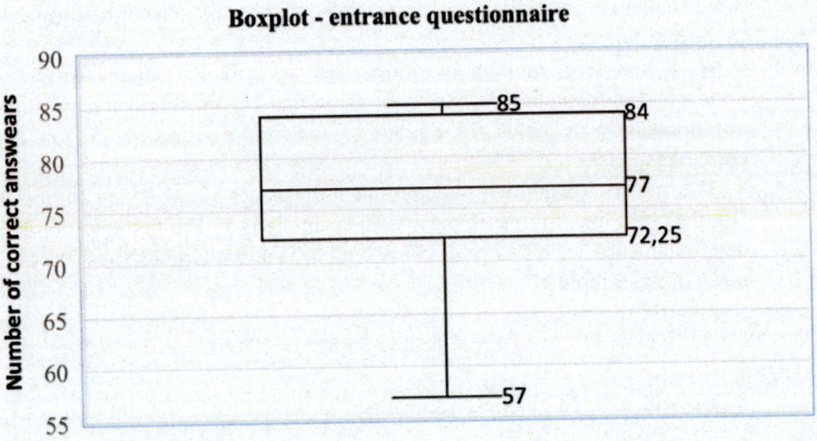

Fig. 14 Boxplot of entrance questionnaire results

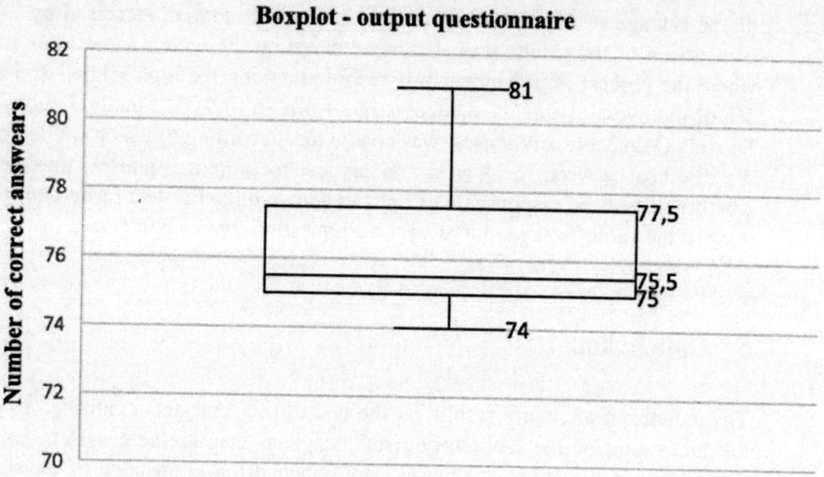

Fig. 15 Boxplot of output questionnaire results

$$T(1) = \frac{\overline{x} - x(1)}{s}\sqrt{\frac{n}{n-1}} = \frac{76,5 - 74}{2.588}\cdot\sqrt{\frac{6}{6-1}} = 1.058,$$

$$T_n = \frac{x_n - \overline{x}}{s}\sqrt{\frac{n}{n-1}} = \frac{81 - 76,5}{2.588}\sqrt{\frac{6}{6-1}} = 1.904,$$

$$Q(1) = \frac{X_{(2)-X_{(1)}}}{X_{(n)} - X_{(1)}} = \frac{75 - 74}{81 - 74} = 0.143.$$

4 Discussion

The chapter presents an analysis and statistical evaluation of the teaching process in which Vernier's modern measurement systems were implemented in the education process of courses: physics and exercises in physics, designed for full-time students in the first semester. This chapter compares the results of students from different types of secondary schools on the basis of filling in an entrance questionnaire consisting of 6 questions and an output questionnaire administered at the end of the semester after the application of Vernier's new teaching aids. From the descriptive statistics produced, we can analyze that when comparing the entrance and output questionnaire, there was an overall improvement of the students when the percentage of the maximum number of correct answers increased by 24%. A comparison of the results was also made based on the secondary schools studied, where the highest improvement was recorded among the high school students by 5% of the overall percentage success rate from 86 to 91%. Finally, Grubbs's and Dixon's Q detecting outlier test was conducted, according to which we determined the distance between the lowest and highest measured responses, thus defining whether one of the responses was not too demanding. The test result showed that none of the values was too far away, so all questions were appropriate.

5 Conclusion

The output questionnaire results (at the end of the semester) confirmed to us that the innovation of the laboratory with the given measurement system increased the interest in the study of physics, strengthened the knowledge of physics, and contributed to the improvement of education process of physics at the Faculty of Manufacturing Technologies. In further research on the application of the Vernier measurement system, we would like to show the advantages of using this system not only in physics classes, but also in mechanics and statics education process.

Acknowledgments This work was supported by the Scientific Grant Agency of the Ministry of Education of the Slovak Republic and the Slovak Academy of Sciences, KEGA 002TUKE-4/2019.

References

1. S. Kanim, X. Cid, The demographics of physics education research. Phys. Rev. Phys. Educ. Res. **16** (2017). https://doi.org/10.1103/PhysRevPhysEducRes.16.020106
2. D. Treagust, R. Duit, H. Fischer (eds.), *Multiple Representations in Physics Education* (2017). https://doi.org/10.1007/978-3-319-58914-5
3. H.Y. Suhendi, M. Ramdhani, S. Irwansyah, Ferli., Verification concept of assessment for physics education student learning outcome. Int. J. Eng. Technol. **7**, 321–325 (2018). https://doi.org/10.14419/ijet.v7i3.21.17181
4. M. Hodosyova, J. Útla, M. Vanyová, P. Vnuková, V. Lapitková, The development of science process skills in physics education. Procedia Social Behav. Sci. **186**, 982–989 (2015). https://doi.org/10.1016/j.sbspro.2015.04.184
5. https://kphv.fvt.tuke.sk/fotogaleria/51
6. D. Darmaji, D. Kurniawan, A. Suryani, Effectiveness of basic physics II practicum guidelines based on science process skills. Jurnal Ilmu Pendidikan Fisika **4**, 1–7 (2019). https://doi.org/10.26737/jipf.v4i1.693
7. E. Etkina, B. Gregorcic, S. Vokos, Organizing physics teacher professional education around productive habit development: a way to meet reform challenges. Phys. Rev. Special Topics Phys. Educ. Res. **13**, 010107 (2017). https://doi.org/10.1103/PhysRevPhysEducRes.13.010107
8. S. Maulidah, E. Prima, Using physics education technology as virtual laboratory in learning waves and sounds. J. Sci. Learn. **1**, 116–121 (2018). https://doi.org/10.17509/jsl.v1i3.11797

9. G. Cebrián, R. Palau, J. Mogas, The smart classroom as a means to the development of ESD methodologies. Sustainability **12**, 3010 (2020). https://doi.org/10.3390/su12073010
10. J.-M. Liang, S. Wei-Cheng, Y.-L. Chen, W. Shih-Lin, J.-J. Chen, Smart interactive education system based on wearable devices. Sensors **19**(15), 3260 (2019). https://doi.org/10.3390/s19153260
11. A. Kenneth, G. John, B. Clarence, V. David, *Physics with Vernier*, 4th edn. (Vernier Software & Technology, Beaverton, 2019)., ISBN 978-1-929075-70-6
12. https://www.vernier.com/product/dynamics-cart-and-track-system-with-go-direct-sensor-cart/
13. A. Miriam, *Štatistické metódy v praxi* (EQUILIBRIA. s.r.o., Košice, 2016)., ISBN 975-80-553-3078-5. https://www.sjf.tuke.sk/kamai/images/vyuka/literatura/statisticke_metody.pdf

Part V
Sustainable Industrial Development

ICT Cost Impact Model in Construction Project Management

Tomáš Mandičák, Peter Mésároš, Annamária Behúnová, and Lucia Zemanová

1 Introduction

Building management is the same as in other industries, i.e., the summary of processes, activities, and activities that execute managers to achieve a set objective. In the case of the construction industry, it amazes a particular building object or building. Management is an accumulated and logically arranged set of knowledge on the principles, methods, and management procedures developed on the basis of abstraction. Management of construction projects means effective integration and coordination of resources to achieve a specified goal.

According to Kerzner [1] project management is a set of activities based primarily on planning, managing, organizing, and controlling enterprises or organizational resources with short-term goals that have been intended to achieve specific objectives and intentions.

Construction project is closely related to investment activity. In the event of a construction project as an investment process, the investment process of construction can be divided into the following phases [2]:

- Pre-investment phase
- Investment phase
- Phase of use and evaluation

T. Mandičák (✉) · P. Mésároš · A. Behúnová · L. Zemanová
Technical University of Košice, Košice, Slovakia
e-mail: tomas.mandicak@tuke.sk

L. Knapčíková, D. Peraković (eds.), *6th EAI International Conference on Management of Manufacturing Systems*, EAI/Springer Innovations in Communication and Computing, https://doi.org/10.1007/978-3-030-96314-9_15

Another view of the construction project phase is described by Všetečka [3]:

- Predesign stage
- Design stage
- Construction stage

Depending on the individual phases, the construction project cost can also be memorised at the selected time period [4].

According to Petráková [4] the construction project costs include all the expenditure of the investor or contracting authority of the construction project, which are related to project preparation, by implementing and submitting a project to use.

The success of CP depends on multiple factors. The intensity and the size of the effect of individual factors for the success of construction projects are different [5]. For a detailed analysis of these success factors, the identification and specification of factors and the rate and intensity of their impact are essential. Chan [6] states human resources, external environment, and communication technologies as key factors for successful technology construction projects.

For the last decades, several research teams and individual authors independently developed the efforts to define critical success factors in managing construction projects. Li [7] identified these factors as effective procurement, ability to implement a project, government security, and so on. Abraham [8] has identified seven factors that affect the CP success.

Based on several studies, it can be said that one of the key factors to successfully control the construction project is the use of information-communication technologies. Mahammad and Saquid have identified other factors and information-communication technologies. Other authors such as El-Saboni, Aouad, Sabani, and Salleh have reported communication systems as a key factor in managing and implementing successful construction projects [9, 10]. It also highlights the importance of ICTs in the process of managing construction projects.

Norizam and Malek [11, 12] defined a successful building project as a project in which all the construction participants (construction project) are satisfied, while this project is implemented without more serious shortcomings, while maintaining the timetable and the envisaged budget.

The exchange of information between the project participants, or within the construction company, can be carried out in several ways. Many construction companies that reject the implementation of information-communication technologies use the traditional method for exchange of information. Communication systems are similar to information systems and technology processes, tools, software equipment, platforms, and people carrying out data exchange, i.e., communication. Some studies point to the great benefits of construction procurement systems. The use of this technology has a particular impact on cost savings [12].

Systems and technologies for managing documents and tools ensuring circulation and exchange of documents are a relatively wide group of tools that have multiple important roles specified in the management and documents. Efficient management of a building project applies all available resources and automation and use all progressive technologies to minimize costs for each activity, including the cost of managing and process documentation.

The benefits of using information-communication technologies in construction project management are directly linked to the benefits of digital exchange, storage, and data sharing. The most commonly mentioned benefits of ICT use include [13]:

- Faster way of communicating—within ICT usage communication is much faster, which represents time saving to other activities
- New ICTs allow multiple users to work at the same time
- Better options for supervision and control
- Simple document filtration
- A positive environmental impact
- Minimizing costs

Based on these studies, a research question arises as to how information and communication technologies affect the costs of arranging construction projects. It is also necessary to define which information technologies have the greatest impact. The solution to this problem is to find an answer that will be very important for practice. This research takes into account the situation in the construction industry and is based on the experience of existing project managers. The involvement of foreign companies operating in Slovakia raises the premise of the general use of new research findings. This new knowledge can form the basis for important decisions in the implementation of information and communication technologies in the management of construction projects. The assumption that some technologies have a positive effect on reducing costs is the basis of research. If this idea is not confirmed by research data, it is not possible to build an economic mathematical model.

2 Methodology

2.1 The Aim of Research

The main aim of this research is to analyze the impact of information and communication technologies on the costs of construction projects (CPs). This process is as demanding as the management of CPs. Therefore, in the final part, the ambition is to find an economic-mathematical model of the impact of ICT on the cost of construction projects. This goal stems from a thorough analysis of already implemented work, as well as general knowledge from practice. This research is carried out in the conditions of the Slovak construction industry and takes these facts into account here.

2.2 Data Collection

The questionnaire was the main tool for data acquisition. The questionnaire focused on the use of ICTs and their impact on the prospects of construction projects.

The questionnaire contained several parts. The first part contained basic characteristic issues (construction size, the participant in the CP, legal form, activity, etc.).

The next part focused on specific information and communication technologies and their use. A substantial part of the questions focused on the impact of these technologies on the costs of construction projects. These data are the basis for the analysis and creation of an economic-mathematical model, the so-called cost impact model.

The questionnaire was distributed in Slovakia. Representatives of construction companies (contractors, sub-contractors, designers, and developers) were addressed.

2.3 Research Sample

As mentioned above, construction companies in Slovakia participated in the research. The participants in the CP answered the examined questions. The structure of the research sample is shown in the table and in the graphic representation. It reflects the representation of construction companies by size and also the percentage of participants in the construction project (Fig. 1 and Table 1).

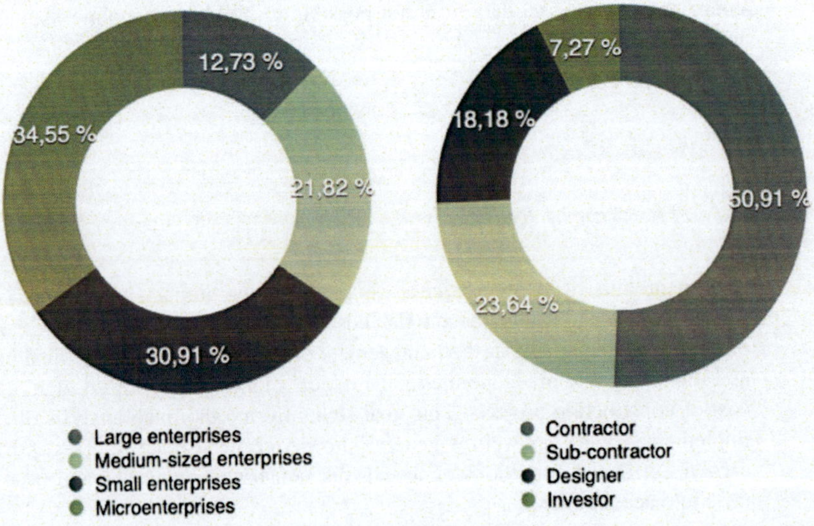

Fig. 1 Research sample by enterprise size and participant of CP

Table 1 Research sample by enterprise size in numbers and %

Research groups	Numbers	%
Large enterprises	7	12.73
Medium-sized enterprises	12	21.82
Small enterprises	17	30.91
Microenterprises	19	34.55

2.4 Data Processing

Data processing was performed using several statistical methods. In the first place, it was the processing and description of descriptive statistics, where the primary data were identified. Subsequently, contingency tables were compiled, and the data were also divided into the monitored groups. Following these conclusions, several other statistical methods came up, which provided a means of summarizing research results. As the aim of the research was to analyze the impact of selected ICTs and design a mathematical-economic model, the AHP method was used. As it is a mathematical expert method, it seemed to be the most suitable means of fulfilling the research goal. The project managers thus answered the questions where they determined the level in the following statement. This method thoroughly compared these values (so-called each with each). The basis of this was the creation of a matrix and subsequent interpretation into a graphical form, where a summary number was generated representing the impact of the technology.

The AHP method has also been used in the decision-making process in several studies on the construction industry [14]. These were researches to identify key factors in the success of construction projects, identify the degree of risk, and determine the success of implementing various materials and technologies or their impact on the importance of criteria such as economic, environmental, functional, aesthetic, and health aspects. The AHP method has also been used to identify and quantify risks at China Railway Construction and logistics [15]. The author's studies aimed at solving several criteria, which are given as factors in evaluating the buildings, also addressed this issue. The authors of the study focused on solving several criteria listed as factors in evaluating building reuse and addressed this issue. These are several examples where the AHP method was used to quantify factors, risks, or impact on selected construction activity. Based on these studies, it is assumed that this method may be suitable for this study's purposes and goal.

The quantities RIJ, the relative significances of the criteria, are arranged in a square matrix of relative significances R [14]:

$$R = \begin{pmatrix} r_{11} \; r_{12} \; \dots \; r_{1m} \\ r_{21} \; r_{22} \; \dots \; r_{2m} \\ \dots \qquad \dots \\ r_{m1} \; r_{m2} \; \dots \; r_{mm} \end{pmatrix} \tag{1}$$

The intensity coefficients of the relation of the i-th variant to the j-th variant are determined in the same way as Saaty states:

$$w_{ij} = \sqrt[m]{\prod_{k=1}^{m} s_m^j} \quad i = 1, \ldots, m; \quad j = 1, \ldots, n. \tag{2}$$

and the final rating is determined as

$$c_i = \sum_{j=1}^{n} w_{ij} v_j \quad i = 1, \ldots, m. \tag{3}$$

3 Results

This model has identified the areas with the greatest scope for cost savings, which may be better for directing attention and aiming to save costs. This research identified key information and communication technologies affecting supply-chain costs. The individual evaluation of the importance of individual categories through expert evaluations is unique. The benefit is considering the benefits and specifics of the industry by experts who operate in the market.

The above also represents the basic benefits of research for practice and the construction industry. This model is of a recommendatory nature based on the statistical probability of impact. The values given by the final information technology impact model represent the coefficient of the probability of the impact on cost reduction.

However, before the impact of these technologies is evaluated and the final model is proposed, it is necessary to describe from the research the real state of use of information and communication technologies. First of all, the use of these technologies is essential and so it is possible to examine these impacts. The degree of digitization in construction is generally insufficient. The potential offered by current technologies is huge. Utilization can be the first step to achieving cost savings in the management of construction projects.

The level of use of ICT varies in the individual phases of a construction project. Information technology in large companies is used at the level of 3.43 in the predesign stage. On the contrary, in medium and small enterprises it is only at the level of 2.01, which is significantly less (Fig. 2).

In design stage, the situation with the use of information technology is the best. Large companies reached the level of 4.04, which is a high utilization rate. Medium and small companies reached the level of 2.17, which is also more than that in the predesign stage.

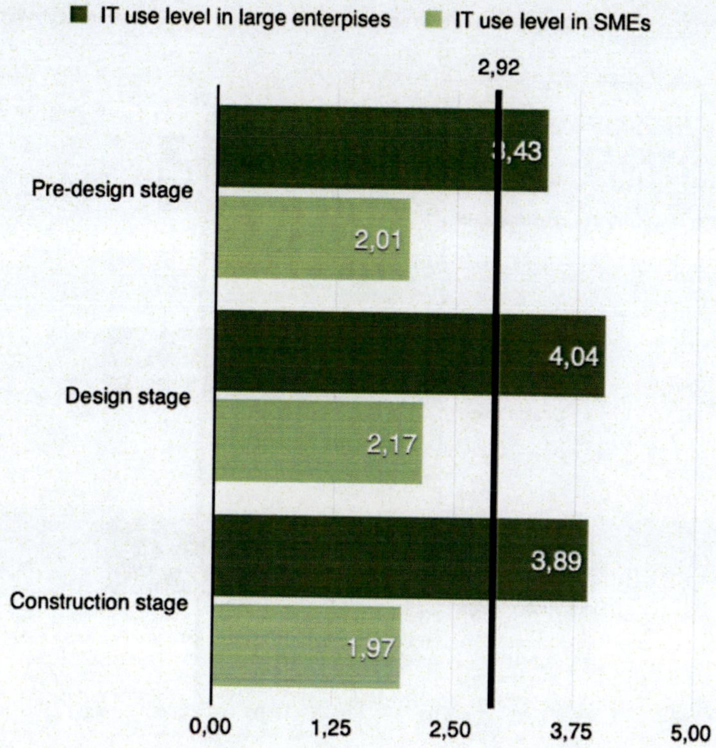

Fig. 2 Information technology use level in each stage of CP

Large companies also reached the level of 3.89 in the construction stage. Small businesses had a value of around 2. The overall average reached 2.92. On this basis, it can be stated that information technologies are used in construction, but their potential is still sufficient for this network to expand.

The situation is similar when using communication technologies. In the pre-design stage, large companies achieved the use of these technologies at the level of 3.28. For small businesses it was 1.75, i.e., significantly less. In the design stage, these results were the best. This means that large enterprises have reached a utilization rate of 3.87 and medium and small enterprises have reached a level of 2.01 (Fig. 3).

The construction stage was also a relative phase for large companies, where communication technologies were used to a large extent, especially for large companies. These values were 3.81 for large companies and 1.77 for medium and small companies.

In the overall assessment, there are also reserves and the potential of today's digital technologies is huge. From this point of view, it should be noted that the

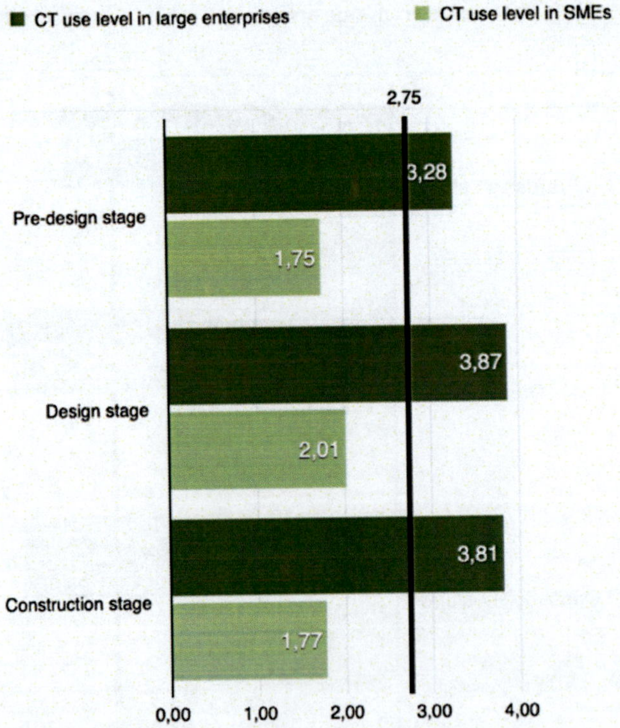

Fig. 3 Communication technology use level in each stage of CP

implementation of advanced technologies has room, especially in small businesses. This is one of the basic findings of this survey.

The information and communication technology impact model's main goal is to reduce a construction project's cost through information and communication technology.

As can be seen from the values, the greatest sense of using information and communication technology is in the predesign stage. In this stage, this model points to the highest coefficient, which represents the greatest impact on cost reduction. This confirms the research hypothesis that information technology's use positively affects costs in individual stages of a construction project.

Based on the AHP method, a model was therefore designed that carries a wealth of information regarding the impact on cost reduction in the management of construction projects (Fig. 4).

This model indicates the degree of impact of technologies in a particular phase of the project. Communication technologies have a greater impact in the predesign phase. In the design stage, they have a significant share in reducing costs, the so-called document management systems. Their strength was also confirmed in the

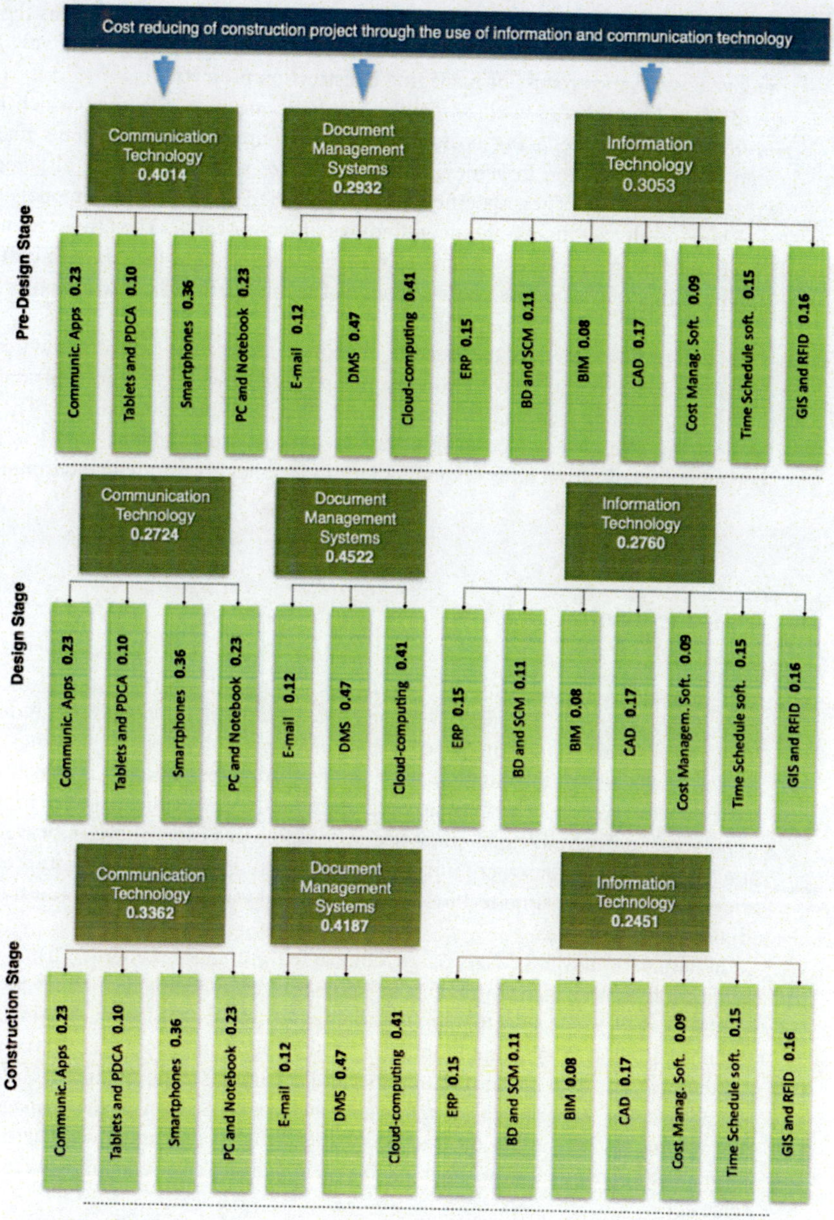

Fig. 4 ICT cost impact model

construction phase. The impact of information technology on cost savings is greatest in the predesign stage.

In terms of specific technologies, the best are cloud systems, as well as the use of smartphones and other communication applications, which will effectively help reduce costs in the process of managing construction projects.

BIM technologies as well as controlling tools are also effective, which have a direct impact on reducing costs. Enterprise resource planning systems make a significant contribution to achieving cost goals. In other words, they also reduce costs by using them. Their specific measure is defined in the proposed model.

These results are among the main findings of this research. The proposed model points out the possibilities of how to effectively contribute to reducing costs by implementing and using individual groups of technologies in construction project management. This is assessed at a very good level.

These results point to the benefit of using information and communication technologies in every phase of a construction project. However, it is good to continue to research, ideally to expand research abroad. The data obtained can be very helpful in practice. The issue of costs is always a topical issue and this area has great potential not only from an economic point of view, but also from an environmental or social point of view.

4 Conclusion

Research aimed at analyzing the current state of use of information and communication technologies and identifying and quantifying the impact on cost reduction has yielded several interesting, and in some cases possibly expected, findings.

The rate of use of information and communication technologies in construction has room for growth. The current situation is not bad, but there is real room for improvement. The implementation of new technologies can achieve improvements at every stage of a construction project. The best results in the use of information alleys and communication technologies are in the design stage. This applies to both groups of technologies.

In terms of business size, large companies are much better at using communication as well as information technology. Large companies achieved better results at every stage, whether it was communication technologies or information technologies. It is the potential for better use that represents room for cost reduction.

Research has also confirmed that the level of implementation and use of these technologies is the basis for better results and cost reduction in the management of construction projects. Construction results, where companies with better ICT equipment operated, achieved a higher savings rate and cost reductions were really achieved.

The proposed model also quantified the individual groups, and how it is possible to achieve cost reduction in the management of construction projects due to the implementation and use of ICT.

These results show, based on the experience and the real situation in the construction industry in Slovakia, how the individual technologies affect cost reduction. Also, based on the model, there is an assumption of how better results in cost reduction can be achieved. However, it should be emphasized that the model is not a guarantee for cost reduction. As for the limits of this research, based on the research sample used, this model reflects the situation in the Slovak construction industry. Despite the participation of foreign construction companies through the representation of branches carrying out construction activities in Slovakia, this model is based on the regular principle. There is also room for further research activities in this area.

The aim is to move this research to other construction markets and extend the research sample to other countries, such as Croatia, the Czech Republic, and Slovakia, where the size of the construction market is comparable to the Slovak one. Also, in the further direction of research, there is the potential to obtain data from Poland or Hungary, which operate in a regional vicinity, and the results should be evaluated and compared. Another goal is to expand the spread and obtain data from Spain. Research has the potential to be extended and applicable to other countries, making its significance global, not just regional.

Acknowledgements This work was supported by the Slovak Research and Development Agency under the contract no. APVV-17-0549.

The chapter presents partial research results of the project VEGA 1/0828/17 "Research and application of knowledge-based systems for modelling cost and economic parameters in Building Information Modelling." It also presents partial research results of the project KEGA 059TUKE-4/2019 "M-learning tool for intelligent modeling of building site parameters in a mixed reality environment."

This chapter presents partial research results of the project VEGA 1/0557/18 Research and development of process and product innovations of modern methods of construction in the context of the Industry 4.0 principles.

References

1. H. Kerzner, *Project Management: A Systems Approach to Planning, Scheduling and Controlling*, 8th edn. (Wiley, New York, 2003)
2. M. Kozlovská, T. Šoltés, L. Sabol, *Plánovanie a riadenie realizácie stavebných projektov* (Technická Univerzita, Košice, 2013)
3. P. Všetečka, L. Belan, *Projektový manažment—I* (Akadémia ozbrojených síl, Liptovský Mikuláš, 2006)
4. Z. Petráková et al., *Využitie projektového riadenia v stavebníctve a investičnej výstavbe* (Vydavateľstvo STU, Bratislave, 2003)., ISBN 80-227-1826-2
5. V. Sanvido, F. Grobler, Critical success factors for construction projects. Constr. Eng. Manag. **118**, 94–111 (1992)
6. P.W. Chan, et al., *A Taxonomy of Knowledge Leakage: Some Early Developments*, in Proceedings of the 22nd Annual ARCOM Conference, Birmingham, vol 2 (2002), pp. 851–861
7. B.A. Li, Critical success factors for PPP/PFI projects in the UK construction industry. Constr. Manag. Econ. **23**, 459–471 (2011)
8. G.L. Abraham, *Critical Success Factors for the Construction Industry*, in Construction Research Congress (2004), pp. 1-9

9. M.A. El-Haram, M.W. Homer, Factors affecting housing maintenance cost. J. Qual. Maint. Eng. **8**(2), 115–123 (2002)
10. A.E. Sekou, Promoting the use of ICT in the construction industry: Assessing the factors hindering usage by building contractors in Ghana, Thesis., http://citeseerx.ist.psu.edu/viewdoc/download?doi=10.1.1.940.7875&rep=rep1&type=pdf, last accessed 2021/05/05
11. A. Norizam, M.A. Malek, Developing critical success factors (CSFs) for effective construction management in Malaysia industry. Asian Soc. Sci. **9**(9) (2013). https://doi.org/10.5539/ass.v9n9p211
12. B.L. Tanko, F. Abdullah, Z.M. Ramly, Benefits of adopting value management to mitigate the problems in the Nigerian Construction Industry. Adv. Sci. Lett. **24**(5), 3818–2822
13. E. Acar et al., Use of information and communication technologies by small and medium-sized enterprises (SMEs) in building construction. Constr. Manag. Econ. **23**, 713–722 (2005)
14. R. Saaty, *Participant Names and Papers*, in Proceedings of the 10th International Symposium on the Analytic Hierarchy/Network Process (ISAHP, Pittsburgh, Pennsylvania, 2009)
15. T.L. Saaty, K. Peniwati, *Group Decision Making: Drawing out and Reconciling Differences* (RWS Publications, Pittsburgh, PA, 2008). ISBN 978-1-888603-08-8

Relationship Marketing: A Modern Marketing Strategy as a Tool to Increase the Competitiveness of the Company in the Market

Annamária Behúnová, Marcel Behún, Lucia Knapčíková, and Lucia Zemanová

1 Introduction

Marketing, nowadays often inflected not only in connection with business, is still largely unknown to many people. We can deduce the meaning of this term from its very name. However, it is not just a matter of selling and buying as it is, but this activity includes wholesaler processes as well as other activities within the market, the aim of which are satisfied customers.

You could say that marketing is as old as civilization itself, but it was not about the marketing we know today. As there was no advertising at that time, trading was based solely on verbal communication. The trader had to learn to convince the buyer and sell his/her product to him/her faster than a competing trader because only the most capable survived in the world of trade. The emergence of book printing had a great impact on marketing, which enabled it to progress. Thanks to the first newspapers to be distributed to the general public, companies were able to advertise and promote their products, resulting in the first paid advertisements [1].

The current form of marketing began to take shape around the nineteenth century. Centuries of the industrial revolution, technological but also social changes, when

A. Behúnová (✉) · M. Behún
Institute of Earth Resources, Technical University of Košice, Košice, Slovakia
e-mail: annamaria.behunova@tuke.sk; marcel.behun@tuke.sk

L. Knapčíková
Department of Industrial Engineering and Informatics, Faculty of Manufacturing Technologies with a Seat in Prešov, Technical University of Košice, Prešov, Slovakia
e-mail: lucia.knapcikova@tuke.sk

L. Zemanová
Faculty of Civil Engineering, Technical University of Košice, Košice, Slovakia
e-mail: lucia.zemanova@tuke.sk

L. Knapčíková, D. Peraković (eds.), *6th EAI International Conference on Management of Manufacturing Systems*, EAI/Springer Innovations in Communication and Computing, https://doi.org/10.1007/978-3-030-96314-9_16

satisfying the needs of customers had to be adapted to these social changes. As demand prevailed oversupply at the time, the pressure to promote was not as great as we know it today. We can talk about the orientation of marketing towards production, because it was industrial progress that brought with it mass production, which, of course, also increased the number of buyers. This phenomenon led to increased production competition, so entrepreneurs were forced to improve their sales strategy. Manufacturing companies have begun to realize the need for market and consumer research to be able to organize their production based on this—that is, when, what, why, and how much to produce [2].

Competition in the world of entrepreneurs intensified until about the middle of the twentieth century. Therefore, competitiveness began to be based on increasing sales through marketing techniques. One of them was, for example, the development of a brand that could not only be sold on the market but also be maintained. Thus, attention began to focus not only on the sale of higher quality products themselves but mainly on improving the brands of these products and services. The result was not only an increase in the company's profits but also an expansion, often an improvement in the company's reputation, all thanks to an increase in the awareness of its brand [3].

Not only marketing but especially branding plays an important role in sales. With the right marketing moves, the company can activate people to buy its products or services. However, the role of branding is to encourage customers to be loyal to a particular brand and therefore to the company itself. It is a process of selling and loyalty to the brand in the long run [4].

This component of sales plays a big role because the brand is what connects customers not only with a specific product but also with the company. It is the brand that can determine the position of the product on the market, and thus its long-term sales. Of course, to achieve customer loyalty to the product brand, it is necessary to take into account the experience with the product, and therefore not only the quality of the product or service must be forgotten, but also the friendly approach to brand representation and sales empathy [4, 5].

The aim of this chapter is to describe CRM, its important position in the company's efforts to increase its competitive position in the market, and the implementation proposal in the conditions of a particular company.

1.1 CRM: Customer Relationship Management

There are many theories about marketing. For example, according to the British Chartered Institute of Marketing, marketing is a management process used to identify and meet customer requirements with a profit. Kotler similarly described marketing, according to which marketing consists of creating, communicating, and selling value to the customer. Marketing as such can generally be defined as a business function that balances the interests of the company and its customers [6].

What these definitions have in common is that they emphasize the customer and his/her needs. They also put an end to the fact that satisfying the customer's needs must be profitable for the company. Marketing is therefore understood as one of the basic or key business functions. Marketing is no less important in a company than funds, logistics, or production.

The main task of marketing is not only to acquire new customers, as many managers think, but the main activity should be to build and improve relationships with existing customers. For this activity, there is a marketing theory CRM, customer relationship management, which means customer relationship management. Businesses focus on building relationships with existing customers because the costs incurred to acquire a new customer are much higher than to retain an existing customer. These costs can be up to five times the cost that a business would incur to retain an existing customer. If a company can build trust and loyalty with a customer, it can bring further profit and a good name to the company in the future, which it can use to gain more customers. From practice, we can say that a company that wants to achieve success in the market must take care of its customers and that it does not lose them [7, 8].

It can be said that CRM is a new method that we have seen in companies for a long time and in various forms, but it was not until the early 1990s that it became known. Through theorists, it began to develop to its present form. To find out what the customer will be happy with and be able to manage relationships with him/her, it is necessary to obtain enough information about the customer. The task of CRM is to purposefully use the obtained customer data. However, customer data alone is not enough if the company cannot know how to control it skillfully. Therefore, the main goal of CRM is to ensure customer loyalty.

To this end, several tasks must be fulfilled, including [9]:

- Providing an up-to-date database of information about our customers in terms of up-to-date products or services that we have sold to customers in the past
- Providing this information about the customer to authorized employees who are responsible for communicating with the customer

To make these tasks easier, it is better to set up a separate department within the company that deals with customer relationship management and is also responsible for managing and controlling these tasks. Most customer relationship theorists recommend centrally managed relationship marketing in a business.

According to Harry Wessling, CRM is "customer interaction." This interaction is provided by a total of four interdependent elements of CRM. The quality of each element is very important, as it affects the entire quality of CRM, and these elements include [6]:

- **Human**: employees of the company. It is essential to ensure that employees have good knowledge and skills to be able to meet the needs and requirements of the customer. Unskilled employees can cause damage not only to the customer but also to the company. Therefore, the company needs qualified employees, to whom it should create a space where they can develop their knowledge and skills.

Fig. 1 Elements of CRM [10]

- **Processes**: Communication and customer processing must be given by certain processes that take place internally in the company and their goal is to meet customer requirements in the shortest possible time.
- **Technology**: These are information technologies with which we can better process and sort information even with a larger number of customers. With this technology, we can share customer data, helping us sort, search, and analyze.
- **Customer information**: It is very necessary and important for the successful sale of products or services. It is not enough for a company to be successful in sales if it has information about its customer. It must also be able to use them correctly. It must know when and what its customer's requirements are to be able to respond to these requirements in time. It is a good idea to update the customer information database regularly. It is not enough to have a lot of information about the customer; it must be available to the staff in charge of contact with customers so that they can use it correctly at the right time (Fig. 1).

Within the basic division of CRM, customer relationship management, we know [11]:

1. **Strategic**: focuses on the core of corporate strategy—customer relations. This strategy focuses on the growth and retention of customers from which the company benefits.
2. **Operational**: Francis Buttle defined this level as a level aimed at automating activities that are associated with direct contact with the customer. Various

CRM software programs and applications automate marketing, sales, and service functions. In practice, in almost every CRM software information is collected about communication with the customer from his/her address to solving customer requirements. The content of this information as well as the entire communication can be shared between the individual departments of the company, especially between the marketing, sales, and customer service departments.
3. **Analytical**: focuses on the analysis of information and collected data and their use for other strategic decisions in the company, such as creating a pricing policy and designing a marketing campaign.

The company needs to implement a CRM system for three main reasons—to reduce costs, by which we mean overall streamlining of processes, reduction of operating costs, and increased revenue due to loyalty of customers who repeatedly buy from the company; to increase the turnover of the company; and to increase the company's competitiveness in the market, which the company must increase or worse maintain if a new competitor enters the market, dealing with the same business sector (trying to reverse the outflow of our customers). In this system, the company has the opportunity to obtain from the customer data the information and outputs necessary for the final evaluation of the cost-effectiveness of customers and on this basis can make further strategic decisions regarding the direction of the company [11].

Harry Wessling, in his book Active Customer Relationships with CMR, named the following benefits of using CRM [6]:

- Seamless process of business processes—reducing delays in processing business orders, reducing the cost of managing these processes
- More individual contacts with the customer
- More time per customer—achieving time savings under the same conditions
- We want to differentiate ourselves from the competition
- Greater prestige
- We gain access to information in real time (all departments have the information needed to meet customer requirements at the same time)
- Reliable and fast forecasts—CRM sells management the information they need to predict future market developments
- Communication between marketing, sales, and services
- Increase the effectiveness of teamwork
- Increase employee motivation—thanks to modern technologies needed for the implementation and integration of CMR, employees have better tools to process their tasks related to customer relations

2 Methodology

The implementation of a CRM system in a company involves many activities that precede its implementation. The implementation of a CRM system must be included in the strategic planning of the company. The company will initially determine the basic requirements for the system depending on the business goals it plans to achieve in the future. It is not enough just to buy a suitable information system for CRM and teach your employees to work with it. A poorly managed CRM system may not affect the higher performance of the sales or marketing department, so we must implement the CRM system uniformly for the entire company [8].

There are six main principles to follow when determining a CRM strategy [7]:

1. The process of developing a CRM strategy must be managed within a strategic project.
2. It is necessary to accurately analyze the current situation in the company, based on which strategic decisions will be selected.
3. Strategic projects must be developed by the company's top management and first-line managers must be actively involved in their management.
4. The methodology developed for the CRM strategy must not be too complicated so that implementation is not too complicated.
5. The chosen strategy must be different from the CRM strategy of a competing company.
6. The chosen strategy must also be practically feasible.

2.1 Deployment Process CRM

The separate process of implementing a CRM system in a company consists of several parts. Each author divides these parts differently, but the sequence of activities and their content are always the same (see Fig. 2).

Idea: This is the initial idea that leads to the implementation of CRM. This idea contains the first reasons why a company needs CRM, what it expects from it, and what implementation will bring. Vision precedes strategy. The vision should be conveyed in a unified form, in a unified philosophy, to the entire company in such a way that each employee is acquainted with the new approach to the customer and voluntarily participates in the creation of a new system [6].

The basic concept: Right at the beginning we must evaluate the internal and external environment of the company. For this evaluation, the SWOT analysis of the company's strengths and weaknesses is most often applied. In addition to the SWOT analysis, the company must divide its customers into groups by dividing them according to their value to the company and the value of their needs. Analyses in the company are the main basis for creating a concept for the implementation of CRM [6].

The internal strategic analysis takes into account the strengths and weaknesses related to CRM. The technologies used are very important for this evaluation, and the people who use these technologies. For properly designed CRM tools, it is important to define the company's competition, and also to analyze the way a competing company communicates. The form of CRM that a company creates

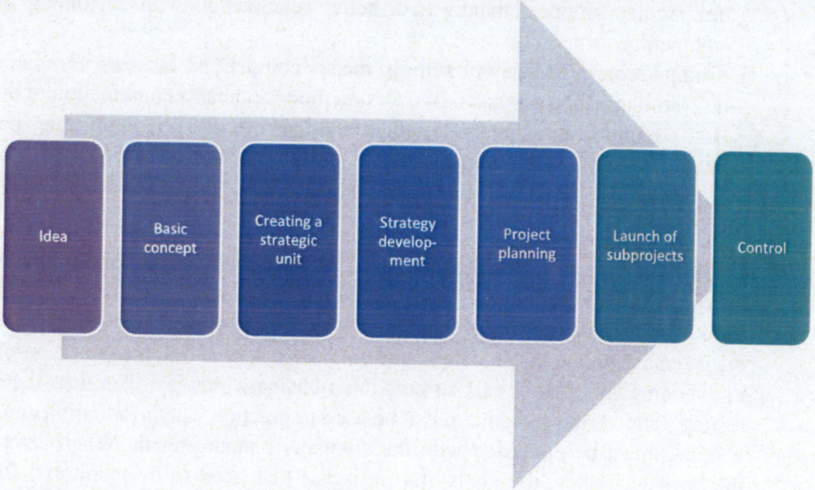

Fig. 2 CRM implementation process [6]

depends on both the subject of the business and the type of communication. The competition offers the customer another option for a better profit from the business relationship. If we allow the competition to the customer to bring innovations and try to bring greater benefits to the customer, there may be a situation where the customer decides to end the relationship with our company and establish a new relationship with a competing company. Precisely because of this situation, it is necessary to closely monitor the competitive environment and analyze its actions in terms of the effect on our customers. The company's competition can be divided into five groups, within which the customer creates a comparative level between individual products and services:

1. Competition in the area of free funds for purchase: competition between products and services that the customer buys for free funds at his/her disposal and for which he/she does not have a predetermined purpose where he/she will use them.
2. Competition in the area of customer problems and their solutions: largely depends on the detailed information that the company has about the customer. A company that has more information about the customer also has a better competitive advantage.
3. Competition in the field of the image: the CRM methodology must be adapted to the image of the company. The company's approach to the customer must be adjusted by the company so that the customer recognizes its brand as easily as possible.
4. Competition for goods and services: if a company does not want to embark on a path of constant price reduction, it can become successful in the market

for similar products thanks to a better relationship with customers than its competitor.

5. Competition in the way of selling: means competition between companies that use similar selling methods. By sales method we mean a combination of business and communication methods, additional services, and, if applicable, payment methods used by the company [6].

Creating a strategic unit of the company: in each company that has implemented a CRM system, a strategic unit should be created, which has the task of managing CRM in the company. Business management sets goals for this unit. In this section, the overall CRM strategy of the company is elaborated in detail into specific tasks for this unit. Development of the strategy by the CRM management unit: the CRM management unit controls other tasks connected with the implementation of CRM in the company. It takes care of creating a business strategy, then tests the created strategy, and verifies whether it can be used in practice. The information and results obtained must be discussed with the company's management. When determining the business policy, the goals and measures that need to be monitored from the point of view of CRM are determined. The given goals and measures are achieved utilizing appropriate CRM tools. The most common of these goals may include increasing customer loyalty, increasing customer value, and increasing customer satisfaction. The settings and possibilities of achieving these goals must be tested and the combination of individual measures verified [6].

Project planning for the implementation of CRM: there is a redistribution of the necessary measures to other projects, which leads to the achievement of set goals, and the CRM strategy begins to take concrete form. In such planning, the CRM control unit must work closely with the individual departments of the company, in particular with the sales department and the marketing department. At this point, important details of the whole strategy are compiled and summarized, which includes resources, costs and time fund, responsible persons, implementation methodology, and project organization. The budget of necessary costs is compiled and the whole plan is approved by the company's management [6].

Launch of CRM sub-projects: at this stage, the final proposals for measures and project launches are implemented, which lead to the implementation of CRM. Because during the implementation of CRM there is a restructuring of the company and changes in various activities, it is, therefore, appropriate if the management unit is part of the top management in the company [6].

Control within the CRM system: after the successful implementation of the CRM system in the company, the company must continuously monitor and evaluate the success of the CRM system. It must also check the effectiveness of the measures used. Only in practice can we determine the quality of the proposed solutions and in the case of detailed monitoring of the process, it is possible to identify its shortcomings, then eliminate them, and bring the system to the desired state. Finding feedback is an important part of any project activity we carry out. As far as CRM is concerned, we perform measurement and verification of the results of CRM implementation to express the profitability of individual sub-processes that are part of CRM. We can apply the obtained measurement results in further streamlining the CRM system process [6].

There are many ways to measure the performance of a CRM system. Because the introduction of CRM brings extensive changes in the company, the measurement of results affects all parts of the company, from operational activities to the company's strategies. The evaluation must always be based on the main objectives we have identified in the CRM strategy. For the measurement itself, we can use one of the general measurement methods, or we can create our measurement method, where we can create our criteria for measuring the success of CRM [12].

Among the generally accepted methods for measuring the success of an implemented CRM, value tools and complex methods of measuring the company's performance are used, such as the balanced scorecard, the EFQM model, or the TQM. As there has been a recent increase in the interest in CRM systems, experts have developed methods aimed directly at measuring CRM, such as the CRM BodyCheck or CRM-CRACK model. Whether a company chooses its method or whether the method already exists, it will always have to adapt this method to its criteria [13, 14].

3 Results

The company in which the research was carried out will focus its business activities on the provision of advertising and marketing services. The greatest effort of the company's management is placed on the sale of advertising space available to the company. Communication with the customer is therefore very important, as competition in this segment is still growing. Currently, the main competitor for print advertising is Internet advertising. As a result, the company begins to feel a reduced demand for their services and begins to show a decrease in advertising revenue. This economic decline has been recorded in the last 2 years. The response to this negative state comes in the implementation of the CRM system.

The company divides its customers into two main groups. Of course, a different way of communicating with the customer also applies to these groups:

1. Existing customers: customers who use the services of the company
2. New customers: customers who have not yet purchased any service

Revenues from advertising customers make up 50% of the company's total revenues. The users to whom the advertisement is distributed do not bring any revenue to the company, but the total number of users is an important attribute—potential customers, who then decide in which printed medium to invest.

The company leases advertising space by addressing customers through sales representatives who meet customers in person. It is one of the most expensive forms of sales. Sales representatives represent up to 75% of the total number of employees in the company. The products that the company prints are distributed to users free of

Table 1 Economic result for 2019 and 2020

	Year 2019	Year 2020	Difference	Difference
Revenues from the sale of advertising space	61,256 €	45,788 €	15,468 €	−25%
Advertising production and distribution costs	12,153 €	11,958 €	195 €	−1.6%
Company overhead costs	31,955 €	29,566 €	2389 €	−7.4%
Profit	**17,148 €**	**4264 €**	**12,844 €**	**−74.9%**

charge, always on the first of the month. Distribution takes place by three employees to all apartment buildings where the company's banners are installed.

There is no CRM system in place in the selected company. However, the company has good prospects for implementing this system. As the company has a wide scope in several directions to diverse customers, it is necessary to implement a CRM system in this company and pay more attention to customer relationship management, which will increase the company's revenue from the rental of advertising space.

Since 2016, the company has reported a decline in revenues from the rental of advertising space. For the decrease in sales not to hurt the company's financial results, its value should be equal to the decrease in costs associated with this activity. This scenario will not be applicable in this case, as the company, even if it is unable to sell a sufficient amount of advertising space, has to bear the costs of printing new advertising, costs associated with exchanging advertising in apartment buildings, remuneration for apartment buildings in which advertising banners are placed, fuel costs, costs associated with renting premises, and, last but not least, wage costs of its employees (see Table 1).

For the company to increase its sales again, it is necessary to perform overall control of the company's costs and try to point out the possibilities of reducing them and at the same time work to increase the company's sales. We will achieve more efficient management of the financial resources available to the company by implementing a CRM system, a suitable strategy, and improving customer relationship management.

To successfully implement CRM in a company, implementation depends on four factors—processes, people, technology, and data. The current value of these factors can be determined either by internal or by external analysis, which is also the basic concept in creating a CRM strategy for the company. We will use this analysis for an effective CRM system solution in the company.

Processes: In the company are one of the main prerequisites for a successful transition to relationship marketing and the implementation of a functioning CRM concept. Therefore, we need to elaborate in detail on the currently functioning processes in the company and, if necessary, propose their modification, which will achieve better system efficiency. In our case, the company does not have processed process procedures, which means that the activities of individual departments are not managed centrally. The activities performed by individual departments are not managed but rather used.

We will take a closer look at the sales department, as this is the most important department of the company. Its activity has a direct impact on the number of sales. This department includes sales representatives who have the task of personal contact with customers. The sales department also includes a call center, whose task is to contact a new client by phone and arrange a meeting date for sales representatives, as well as provide support for the sales department with a line for customers.

The sales department is further divided into other sections:

- Communication and advertising section: This section has the task of preparing a campaign and determines how new products are communicated to customers (both user and customer).
- Marketing department for the customer: The task of this department is to work on the preparation of new products for the customer, followed by the elaboration of price offers and sales aids for the sales department. The marketing department also provides market research and analysis focused on the competition, which deals with advertising and rental of advertising space in the area.
- Marketing section for users: This section deals with the improvement of services provided by the company. It also has the task of increasing interest in the service from the user's point of view.

3.1 Company Strategy

The overall direction and strategy of the examined company are created by the management led by its owner. The creation of the strategy depends on various surveys of competing media and also on users, in this case, residents of apartment buildings. The company places insufficient emphasis on internal communication. To create a strategy, it takes into account internal factors such as finance and budget.

Information on which company has chosen a new strategy is often insufficiently presented within the company's departments, and based on verification, we found that information on the objectives of the new strategy was insufficiently communicated to departments, with the result that individual departments were not informed of the more detailed objectives of this strategy. If we want to introduce new changes or a new company strategy in the company, we must also focus on ensuring that this strategy is communicated sufficiently and clearly to the relevant departments and managers to be committed to meeting its goals. This will also apply to the implementation of the CRM system and its subsequent notification to individual departments. Otherwise, a lack of information can cause the company many serious problems.

3.1.1 Information Technologies

For the successful implementation of a CRM system in a company, the company must be equipped with its information system. The analyzed company already has an information system in which all data about its customers are registered, called the PEN system. It is possible to enter data from various sources into the system. By this, we mean manually entering customer data. This data is added to the system mainly by the sales department. A financial platform is also created in the system, which serves the economic department for subsequent invoicing and monitoring of receivables. We evaluate the processing of documents positively. All documents are prepared in advance in a template (this applies to contracts and orders). The sales representative only changes the customer's business data according to the client card; he/she does not have to enter anything manually. As for document archiving, there are still shortcomings—archiving takes place exclusively in paper form, while electronic archiving of documents saves time when searching for specific orders and contracts and facilitates the sharing of information.

3.1.2 Information

Most customer information is obtained directly from customers from signed contracts and orders. This information is only basic. The history of the customer's previous orders is included in the records, but his/her wishes or complaints are not recorded. The data is updated only when the customer orders a new service and signs a new advertising order. The call center is only used for incoming calls. As for telemarketing, it is used only very superficially. Only contacts for customers who have indicated that they are interested in advertising are collected. The call center should be used much more efficiently so that one of its tasks should be to contact customers who have used the company's services in the past.

3.1.3 Employees

The quality of a CRM system is closely related to the employees who use its tools. In the analyzed company, all new employees go through a training process. Sales staff (sales representatives) and call center operators must complete a 2-week course to familiarize themselves with the workflows and products that the company offers on the market. The evaluation of employees takes place based on monthly deadlines, depending on the success of the sale of advertising. Employees and their quality are one of the key components of customer relationship management.

3.1.4 External Analysis

Every year, the company surveys the surrounding market. Perceptions of external factors are very important for the society. It tries to monitor the marketing strategy of the competition and based on that it creates and adapts its strategy. It also monitors the use of its products against the competition. Competitors in the region for our chosen company are advertising newspapers, operators of billboards, and illuminated advertising billboards.

3.2 *Proposal for CRM Implementation in the Company*

Based on a previous analysis of the situation in the company, we can say that the company needs the implementation of CRM and more effective customer relationship management due to declining sales of advertising space and the departure of more customers in recent years.

The vision for the future that the company should address is to retain existing customers who regularly renew the order and strive to gain new customers or to resume cooperation with lost customers. To inform employees about this vision, we recommend personal interviews with employees in meetings with the company's management, in which all information about the new approach to customers will be explained to them.

The main points for creating a CRM system in the company are the following:

- Creation of a new department in the sales department, which will be responsible for managing CRM in the company and evaluating sales plans. This department will be directly influenced by the owner of the company.
- Development of the CRM strategy: the new department of the sales department that we propose to create should set as the main goal the growth of the customer's value and the growth of his/her trust in the company.
- To make the work more efficient, we recommend expanding the PEN system used by the company with the possibility of archiving documents in electronic form.
- Improving the registration of customers who are interested in cooperation.
- Expansion of the call center's activities to address new and lost clients and the introduction of recording of their communication. Also, set up a record of communication between the sales representative and the client.
- Introduction of ordering services via the Internet or an application on a mobile phone, thus enabling the transition from a personal business solution to an electronic one. To a large extent, this relieves trade representatives.
- Continuous control of the company's culture and monitoring the interest of employees in fulfilling the company's goals.

The newly created section must still determine in practice the fulfilment of the given goals. It depends on how much money the company will be able and willing

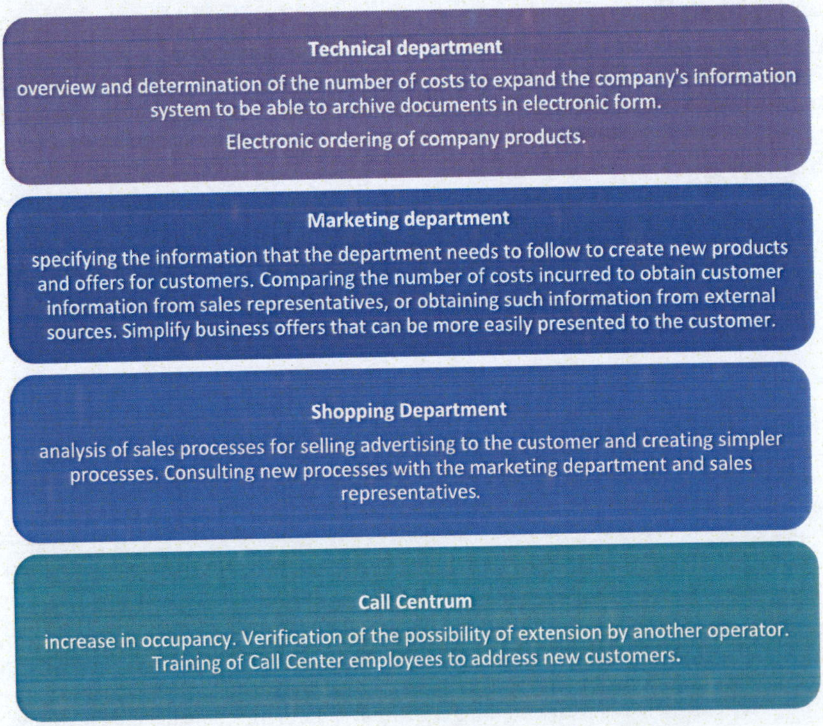

Fig. 3 Implementation of CRM in the company through all departments and divisions

to allocate to change. Planning the implementation of a CRM system in a company is very important and the cooperation of all departments and divisions is necessary to define the schedule, costs, and resources needed for the implementation of the system (see Fig. 3).

4 Conclusion

The introduction of a CRM system in a company is currently an easily acquired tool that gives the company an advantage over the competition, of which there is a large number in today's market. The aim of this chapter was to describe CRM, its important position in the company's efforts to increase its competitive position in the market, and the implementation proposal in the conditions of a particular company. The company in which the research was carried out has a good precondition for implementing a CRM system. The economic situation in the company has not been favorable in recent years, so the introduction of CRM, as one of the solutions, offers an opportunity to improve this unfavorable situation.

As it was found based on internal company documents, the company records a decrease in customers and at the same time is not able to reduce its costs linearly. With the help of the proposed changes, the company has the opportunity to monitor and analyze the processes and ways of communication with customers through which the necessary revenues are lost and the activities that are inefficient for the company. At the same time, it brings proposals for change and more effective communication with customers, which can help companies acquire new as well as their former customers, which will lead to a renewed increase in sales and increase the company's competitiveness in the market.

References

1. M. Spišáková et al., Construction waste audit in the framework of sustainable waste management in construction projects—case study. Buildings **11**(2), 1–16 (2021)
2. M. Blažková, *Marketingové řízení a plánovaní pro mále a střední firmy* (Grada Publishing, Praha, 2007)
3. M. Karlíček et al., *Základy Marketing* (Grada Publishing, 2013)
4. R. Lehtinen, *Aktivní CRM—Řízení vztahů se zákazníky* (Grada Publishing, Praha, 2007)
5. V. Chlebovský, *CRM—Řízení vztahů se zákazníky* (Computer Press, Brno, Brno, 2005)
6. H. Wessling, *Aktivní vztah k zákazníkům pomocí CRM* (Grada Publishing, Praha, 2003)
7. F. Buttle, *Customer Relationship Management: Concept and Tools* (Elsevier, Amsterdam, 2004)
8. Customer think: How do you evaluate your CRM implementation. https://customerthink.com/
9. L. Ryals, S. Knox, Cross-functional issues in the implementation of relationship marketing through customer relationship management. Eur. Manag. J. **19**(5), 534–542 (2001)
10. INFOPLEX CUSTOMER RELATIONSHIP MANAGEMENT CRM SOFTWARE? https://techpalace.com.ng/product/infoplex-customer-relationship-management-crm-software/
11. J. Galbreath, Relationship management environments. Credit World **87**(2), 14–22 (1998)
12. P. Němček et al., *Možné Trendy rozvoje podniků* (Akademické nakladatelství CERN, Brno, 2004)
13. P. Mésároš et al., Adaptive design of formworks for building renovation considering the sustainability of construction in BIM environment-case study. Sustainability **13**(2), 1–20 (2021)
14. P. Kotler, *Moderní Marketing* (Grada, Praha, 2007)

Using Economical-Mathematics Method of Input-Output Analysis as a Tool for Cost Calculation of Steel Construction Production

Matouš Machka and Theodor Beran

1 Introduction

The idea of implementing modern methods of managerial accounting in small and medium-sized industrial enterprises is currently a very appealing topic. Its attractivity is also compounded by the fact that these enterprises must create a strategy that will enable them to secure and maintain a stable position in the market. To meet this target, small and medium-sized enterprises must face many challenges. Among the most fundamental are, for example, expanding their production portfolio, introducing new production technologies, training employees, and increasing the quality and safety of the working environment [1]. In order to accomplish these purposes, it is obvious that we need modern cost accounting methods that will enable more efficient cost management and planning in comparison with traditional methods [2]. However, the implementation of such methods in small and medium-sized enterprises faces an array of obstructions [1, 2]. The existence of such barriers prevents small and medium-sized enterprises from applying such methods, not only in the Czech Republic, but also in the whole world [1, 3, 4]. The demands of current practice are impelling engineers to design new methods that would ease the difficult implementation of modern cost management methods in small and medium-sized enterprises. The processing of such appropriate methods strongly depends on the type of the particular company, and requires a detailed analysis of its organization and production activity [1].

The authors of this chapter have attempted to create a new cost management method based on the input-output model, supplemented by a set of cost-incurring activities. The new methodology should strengthen planning and cost management

M. Machka (✉) · T. Beran
Czech Technical University in Prague, Prague, Czech Republic
e-mail: Matous.Machka@fs.cvut.cz

L. Knapčíková, D. Peraković (eds.), *6th EAI International Conference on Management of Manufacturing Systems*, EAI/Springer Innovations in Communication and Computing, https://doi.org/10.1007/978-3-030-96314-9_17

within the main production activity of the company Jetti [5]. One of the main production activities is the custom manufacture of steel constructions. This work follows on from a previous publication [6] in which the first proposal of this methodology relating to the production of steel constructions was put forward. In the following sections, this proposal will be tested in detail in the calculation of costs for the assembly parts of the notional steel construction. The concept for the steel construction was created based on commonly produced steel constructions, and perfectly matches their design and technological complexity.

2 Design Analysis of the Steel Construction

To establish the input-output model, it is essential to have a detailed specification of the individual components and assembly groups of the given assembly, i.e., the steel construction assembly in our case. Steel constructions are mainly formed by a large number of regularly repeating assembly groups plus other components primarily used for bracing. It is also important to note that while many types of identical assembly parts may be found in several different assembly groups, they may also be necessary for the final assembly of the steel construction itself. A good basis on which to form the input-output model is to find the relations and connections among the respective components and assembly groups. To identify the natural relations properly, the drawing documentation included in the technical report must be analyzed in detail, especially the design part. However, such documentation is not always ideal, and so the following section addresses this issue.

2.1 *Analyzing the Technical Report*

The implementation of the input-output model as a tool for cost management, especially in small and medium-sized enterprises, is generally beset by multiple obstructions [7, 8]. The basic and all-important hurdle to overcome is selecting suitable values for the first quadrant of the input-output model [7], i.e., correctly setting the technical coefficients. The initial source of these values derives from the bill of materials (BOM) for components and assembly groups attached to the design part of the technical report.

The essential problem lies in the structuring of each BOM. Small and medium-sized industrial enterprises with custom manufacturing as their main production activity receive already processed technical reports together with the order. The quality of technical reports, especially the chapters dealing with construction, is not suitable in most cases for a high-quality input-output model. The input-output analysis concept is based first and foremost on directly observable phenomena characterized by interdependences [9, 10]. To be able to apply the input-output analysis in our case, first we have to appropriately arrange the initial data necessary for constructing the final input-output model.

The main disadvantage of such externally managed reports comes in the illogical sequence of components and assembly groups in the bill of materials that is completely contrary to the principles and rules of design engineering.

Such a chaotically formulated bill of materials not only impedes the start of the production process, but it also fails to create the conditions in which an input-output analysis could be applied. The cause of such inconsistencies lies with the staff of these external companies, i.e., customers. The bill of materials is often processed by novices, employees who lack sufficient experience or sufficient education in this area. The result includes delays in the main production activity due to deciphering a poor BOM, and in some cases it means drawing up a new BOM from scratch.

A proposal to improve the current situation is presented in the following subsection, where a new BOM format relating to the notional steel construction is described. This bill of materials has been formulated to fully meet the essential conditions for implementing the input-output modelling.

2.2 *Bill of Materials Structure Based on Input-Output Model Requirements*

The basic prerequisite for creating meaningful drawing documentation is dividing the drawings into coherent groups. Here, we have decided for a three-stage division (see Table 1) where the steel construction is denoted as OK, and the XX and Y symbols represent numerical variables.

This three-stage system provides a comprehensive overview of the respective production components and assembly groups; however, it is not sufficient for input-output modelling. Therefore, the respective assembly parts will be further divided into three class groups.

The first two class groups are designated for separate production components and semifinished products, and the third class is designated for assembly groups. With respect to a large amount of assembly parts, we have to further classify these class groups into individual classes that will be denoted with lower case letters and corresponding indexes. The criteria for this index classification are stated in Table 2. By these criteria we basically mean conditions for assigning an index to the respective class belonging to the given class group. The classes belonging to Group

Table 1 Division of drawings into three groups in the drawing documentation

Drawing no. format	Specification	Quantity of drawings
OK—0—00—XX	Production drawing of individual parts	25
OK—Y—XX	Assembly group drawing	32
OK—1	Assembly drawing	1

Table 2 Classifying the respective components and assembly groups for input-output analysis

Class group	Definition of class groups of assembly parts	Criteria for index classification
S	Separate components of sheet metal	Material, production technology
H	Separate components of steel beam profiles	Semifinished product, production technology
M	Assembly group	Purpose of the group in the assembly, technology

Table 3 List of drawings of basic components made of hot-rolled sheet metal belonging to class Group S

Material	Thickness [mm]	Drawing no.	Class	Component specification	Quantity
S355	20	OK—0—00—01	s_1	Column foot	12
S235	12	OK—0—00—02	s_2	Bracing attachment	12
		OK—0—00—03			7
		OK—0—00—04			11
		OK—0—00—05			17
		OK—0—00—06			2
		OK—0—00—07		Static bracing attachment	12
		OK—0—00—08	s_3		4
		OK—0—00—09	s_2	Attachment of IPE support profiles	44
		OK—0—00—10		Support bracket face	76
		OK—0—00—11		Dynamic bracing attachment	24
		OK—0—00—12			6
		OK—0—00—13			2

S and Group H have only one index that is assigned based on the design similarity of the respective components, but also based on the same production technology. For classes belonging to Group M, we designate two indexes. The first index expresses the purpose of the given assembly group in the final assembly, and the second index depends on the quantity of components used for the given assembly group.

We will use this division not only for designing the input-output model, but also for managing production operations and calculating phase costs where the cost objects will represent index groups.

This index classification is applied in full in Tables 3, 4, and 5. It is obvious that this classification may be implemented only when we have clearly apprehensible drawing documentation at our disposal. Mathematically, the respective class groups may be expressed as (1)

$$M = f\,(S, H) \tag{1}$$

Table 4 List of drawings of components and standardized semifinished products made of hot-rolled steel profiles, with material quality of S235JR, belonging to class Group H

Semifinished product	Length [mm]	Component drawing no.	Class	Component specification	Quantity
HEB 220	3800		h_1	Basic vertical beam	12
HEA 220	356.6	OK—0—00—14	h_2	Transverse and longitudinal horizontal beams	28
	506.5	OK—0—00—15			5
	Various		h_3		30
IPE 200	480	OK—0—00—16	h_4	Transverse bracing	3
	524	OK—0—00—17			2
	1554	OK—0—00—18			7
	1634	OK—0—00—19			9
Square tube 80 × 80 × 5	Various		h_5	Steel beam profiles for bracing	12
Square tube 150 × 150 × 4	1600	OK—0—00—20	h_6	Inclined beam	8
L 60 × 60 × 6	500	OK—0—00—21	h_7	Side bracing	11
	590	OK—0—00—22			8
	650	OK—0—00—23			6
	665	OK—0—00—24			8
	690	OK—0—00—25			2

This relation expresses the dependency of the assembly groups on basic components made of sheet metal and steel beam profiles, i.e., each assembly group consists of these types of components. After this classification, the production process may begin.

3 Technology Analysis of the Steel Construction

This chapter focuses on the arrangement of the production process within the respective production departments of Jetti. A thorough knowledge of the production process will help us better understand the relations between the company's main and the secondary production operations, especially with respect to the capacity load of the respective in-house manufacturing units. The production process itself is divided into four phases that are depicted in Fig. 1. The aim of the technology analysis is to set up the production process to ensure continuous operation between the respective phases without any major interference or conflict that would adversely affect the final cost calculation. A high-quality technology analysis contains a description of the activities directly related to the production process; moreover, it provides a basis for applying the modern cost accounting methods ABM/C (activity-based management/costing), which is directly related to the cost-incurring operations.

Table 5 List of assembly group drawings falling into class Group M

Drawing no.	Class		Assembly group specifications	Quantity
OK—1—01	$m_{1\,1}$	M_1	Supporting columns	1
⋮	⋮			⋮
OK—1—12	$m_{1\,12}$			1
OK—2—01	$m_{2\,1}$	M_2	Center cross	2
OK—3—01	$m_{3\,1}$	M_3	Support brackets	5
OK—3—02	$m_{3\,2}$			4
OK—3—03	$m_{3\,3}$			1
OK—3—04	$m_{3\,4}$			2
OK—3—05	$m_{3\,5}$			1
OK—3—06	$m_{3\,6}$			1
OK—3—07	$m_{4\,7}$	M_4	Support brackets (welded)	1
OK—3—08	$m_{4\,8}$			1
OK—4—01	$m_{5\,1}$	M_5	Auxiliary support brackets	1
OK—4—02	$m_{5\,2}$			1
OK—4—03	$m_{5\,3}$			3
OK—4—04	$m_{5\,4}$			1
OK—4—05	$m_{5\,5}$			1
OK—4—06	$m_{5\,6}$			1
OK—4—07	$m_{5\,7}$			2
OK—5—01	$m_{6\,1}$	M_6	Bracing	3
OK—5—02	$m_{6\,2}$			6
OK—5—03	$m_{6\,3}$			1
OK—5—04	$m_{6\,4}$			2

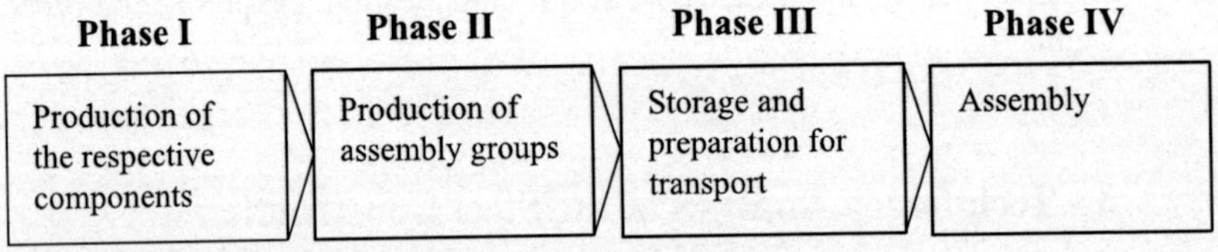

Fig. 1 Schematic representation of the respective production phases of the steel construction. Source: own

For better production management with respect to potential scrap generation, especially with welded assembly groups, it would be advisable to perform a PERT analysis that contains stochastic formulas. A high-quality input for calculating the probability of scrap generation is represented by the experience of production managers, who are able to predict where errors may occur. With these assumptions, we can better predict the duration of operations with respect to possible time allowances for any potential repairs/corrections. If the company employs experienced production staff that are able to predict such time allowances, application of critical path method (CPM) is sufficient.

Jetti has five production buildings, i.e., production departments, where both the main and the secondary production operations take place. Within the internal management, however, we will be working with manufacturing units where the main production operations primarily happen. These manufacturing units have a precisely defined working space, and form part of the respective production buildings. The introduction of manufacturing units is essential mainly in terms of managerial accounting, because records of overhead costs are kept in every manufacturing unit, and these overhead costs are important for the final cost calculation [11].

3.1 *Production Operations, Phase I*

In this phase, production is prepared of the respective components necessary for assembly groups and the steel construction itself. The individual steps of this phase are described in Table 6, where the index classification of the respective classes formulated in the previous chapter is fully implemented.

According to Table 6, we can assume that the biggest initial load relates to manufacturing unit SB, where most of the production operations will be performed. Critical, however, is choosing the sequence of sheet metal semifinished products and steel beam profiles that are intended for processing using the respective

Table 6 Description of the steps necessary for compliance with the requirements of production activities in Phase I

	First step		second step		3rd step	
Class	Technology	Center	Technology	Center	Technology	Center
s_1	Precision laser cutting	SA				
s_2			Edging	SC		
s_3						
h_1	Cutting by CNC band saw	SB			Cleaning areas for welding and final surface treatment	SD
h_2			Cutting using a handheld angle grinder	SB		
h_3						
h_4			Drilling holes for screws	SB		
h_6						
h_5	Cutting using an NC circular saw					
h_7			Drilling holes for screws	SB		

production machines. Due attention must be paid to this issue because this is custom manufacturing, where adherence to the contractually agreed deadline is crucial. To ensure an optimal production run, I propose selecting the sequence stated in Table 6. The reason being is that this proposed sequence fully complies with the class categorization that takes into account the design-technological similarity of the respective manufactured assembly parts. Moreover, we need to consider the fact that every class consists of other parts. This applies especially to class Group H, where individual components are divided based on their length, which is specified for each component in Table 4. The sequence of these components especially depends on their amount, i.e., production will start with the most represented component in the given class.

3.2 *Production Operations, Phase II*

In Phase II, the production of assembly groups consisting of the respective components represented by class Groups S and H is performed. The technological process is identical for all assembly group classes. The first step is to weld together the individual assembly parts according to the specified production drawings stated in Table 5. Welding is executed in manufacturing unit SE. In the second step the welded assembly groups are cleaned and prepared for the final surface treatment. The cleaning of the assembly groups is also carried out in manufacturing unit SD, and the final surface treatment is performed in manufacturing unit SF. A detailed breakdown of the respective components for the specified assembly groups is stated in Table 7. This breakdown is based on the technical report and represents the base for setting the input data into the first quadrant of the input-output model.

3.3 *Production Operations, Phases III and IV*

The third production phase relates to warehousing. Finished assembly groups and individual assembly parts are stored in the special manufacturing unit SG. Storing depends on the sequence of deliveries from the manufacturing unit SF, where both the assembly groups and individual components of classes h_4 and h_7 are painted. Components of this class are necessary for the final assembly, and their quantity is manifested in the input-output model, namely in the second quadrant. After the final deliveries from the manufacturing unit SF arrive, and the last warehousing phases are performed, transportation may begin. Transport is connected with loading finished parts onto trucks, where considerable attention must be paid to the limited truck capacity. The overhead staff of manufacturing unit SG must calculate where best to put all the finished parts in the loading area so that the whole construction can be delivered directly to the assembly destination by company trucks. The logistics

Table 7 Breakdown of the quantities of individual components necessary for building the given assembly group

Class	s_1	s_2	s_3	h_1	h_2	h_3	h_4	h_5	h_6
$m_{1\,1}$	1	4	2	1	1	1	0	0	0
$m_{1\,2}$	1	6	0	1	2	0	0	0	0
$m_{1\,3}$	1	5	0	1	2	2	0	0	1
$m_{1\,4}$	1	5	0	1	2	1	0	0	1
$m_{1\,5}$	1	4	0	1	1	1	0	0	1
$m_{1\,6}$	1	6	2	1	2	1	0	0	1
$m_{1\,7}$	1	7	0	1	2	1	0	0	0
$m_{1\,8}$	1	8	0	1	4	1	0	0	1
$m_{1\,9}$	1	8	0	1	3	1	0	0	1
$m_{1\,10}$	1	5	0	1	2	1	0	0	1
$m_{1\,11}$	1	6	0	1	2	1	0	0	1
$m_{1\,12}$	1	3	0	1	1	1	0	0	0
$m_{2\,1}$	0	12	0	0	4	2	0	0	0
$m_{3\,1}$	0	20	0	0	0	5	0	0	0
$m_{3\,2}$	0	24	0	0	0	4	0	0	0
$m_{3\,3}$	0	5	0	0	0	1	0	0	0
$m_{3\,4}$	0	6	0	0	0	2	0	0	0
$m_{3\,5}$	0	4	0	0	0	1	0	0	0
$m_{3\,6}$	0	3	0	0	0	1	0	0	0
$m_{4\,7}$	0	11	0	0	4	1	0	0	0
$m_{4\,8}$	0	3	0	0	1	1	0	0	0
$m_{5\,1}$	0	1	0	0	0	0	1	0	0
$m_{5\,2}$	0	3	0	0	0	0	1	0	0
$m_{5\,3}$	0	6	0	0	0	0	3	0	0
$m_{5\,4}$	0	3	0	0	0	0	1	0	0
$m_{5\,5}$	0	4	0	0	0	0	1	0	0
$m_{5\,6}$	0	5	0	0	0	0	1	0	0
$m_{5\,7}$	0	4	0	0	0	0	2	0	0
$m_{6\,1}$	0	12	0	0	0	0	0	3	0
$m_{6\,2}$	0	12	0	0	0	0	0	6	0
$m_{6\,3}$	0	4	0	0	0	0	0	1	0
$m_{6\,4}$	0	4	0	0	0	0	0	2	0

management should not be forced to outsource transport, since this activity could also increase fixed costs, in particular.

The final assembly of the construction depends on the contractual conditions. However, the company must not forget about calculating quantities of standardized components, e.g., studs, hexagon screws, nuts, and washers. These items form another important part of direct material costs.

4 Economic Analysis of the Steel Construction Production

During production of the steel construction, mutual cooperation takes place between the respective in-house manufacturing units. To efficiently calculate costs for the individual cost objects represented by the respective classes here, we use the methods of applied input-output analysis. These methods, also referred to as matrix calculations of in-house service costs, are calculations that use mathematical models.

The essential step in applying an economical-mathematics method of input-output analysis is allocating initial costs to the respective cost objects that represent the third quadrant of the input-output model. This is described in detail in the following chapter.

4.1 The Issue of Initial Cost Allocation

When allocating initial costs, we use the information formulated in the section on technology breakdown. That section maintained that the manufacturing process is divided into four phases. The essential phases are the first two, which are also the most expensive. Now, we will describe these two phases from the economical point of view.

Phase I relates to the manufacture of the respective components, and is divided into three steps depending on what class is concerned. Since the respective components are made from the same semifinished products and identical technologies are used for their production, we will first look at the calculation of costs in the respective manufacturing units. For this purpose, we will use the absorption method of the phase cost calculation, where direct material cost is of great significance. The structure of direct material cost items is shown as semifinished products stated in Tables 3 and 4. Of course, direct labor costs must be taken into consideration. In this case it will involve metal workers. The overhead costs reported in the respective manufacturing units must also be included in the calculation. Specifically, the following manufacturing units are concerned: SA, SB, SC, and SD.

Phase II is represented by building the assembly groups. Here, we are no longer counting direct material costs, since this was processed as initial costs in Phase I. And they logically enter Phase II as secondary costs. The crucial role in this case is played by direct labor costs, e.g., welders and painters, including a significant proportion of overhead costs registered in manufacturing units SD and SF. Also here, the absorption method of the phase cost calculation is applied.

4.2 Matrix Cost Calculation

The basic scheme of the input-output model is illustrated in Fig. 2, where individual symbols represent the following: $A_{n \times n}$ square matrix of technical coefficients, $IT_{n \times n}$ square matrix of internal turnover, Y_n column matrix of final use, Z_n column matrix of initial costs and profit, and X_n column matrix of gross turnover. The matrix of technical coefficients is calculated from Relation 2 where x_{ij} represents the respective secondary costs:

$$A_{n \times n} = [a_{ij}] = \left[\frac{x_{ij}}{X_j}\right] \tag{2}$$

Individual input-output model matrixes are located in quadrants. The matrix of internal turnover is in Quadrant I. The matrix of final use is located in Quadrant II, which represents the sum of produced goods and the status change of unfinished production. The matrix of initial costs is located in Quadrant III, which expresses total production. A high-quality base for working with input-output models is provided by [12].

The matrix cost calculation deals with the three basic tasks of input-output modelling [12]. The first task enables us to calculate the useful final product defined by a matrix Y_n from the total production volume. With the second task we are able to calculate the necessary total production volume defined by the matrix X_n based on the given final product. The third task deals with calculating the effect of changes in costs on the pricing of company production. The principles of this task are explained in the following section.

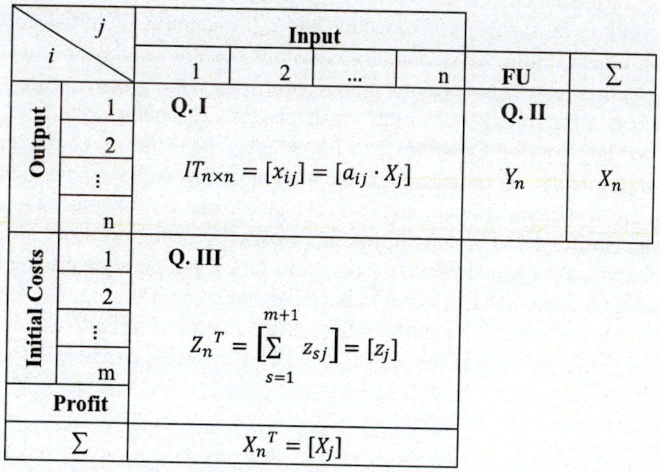

Fig. 2 Standard scheme of the input-output model. Source: own

4.3 Formulating the Third Task of Input-Output Modelling

The point of the pricing task is to calculate unknown price coefficients. If we have an initial cost matrix with given values, we describe it in the standard way Z_n. However, if the values of initial costs change, a new matrix is formed. Write down the new matrix of initial costs and profit with an apostrophe Z'_n. These matrixes already contain totaled initial costs for given cost objects; therefore they have a column shape. In this task, we consider only initial costs without profit. It is obvious that a change in the initial costs also causes a change in the gross turnover. The extent of such change is represented by so-far unknown price coefficients. The pricing task is expressed by the Eq. (3)

$$C_n = \left[E_{n\times n} - A_{n\times n}{}^T\right]^{-1} \bullet W'_n \tag{3}$$

where the respective matrixes represent:

$C_n \ldots$ matrix of unknown price coefficients
$E_{n\times n} \ldots$ identity matrix.

Matrix W'_n included new values of initial cost divided by values of gross turnover. This matrix is defined by the following Relation (4):

$$W'_n = \left[\frac{z'_i}{X_i}\right] \tag{4}$$

The pricing task defined by Relation (3) shows that a change in initial costs changes the price structure of relations in the whole model to which it is applied.

The presented pricing task may be used to calculate the final calculations for individual cost objects if we only know the amount of initial costs for a given cost object. However, secondary costs are not known. In the previous chapter, a theoretical procedure was formulated for ascertaining initial costs for given cost objects represented by assembly part classes. The question is how to replace the secondary costs in Quadrant I, i.e., how to set the technical coefficients. If we have a high-quality information base about the quantity of pieces and their arrangement, we can use them in Quadrants I and II, since every product piece represents a partial calculation cost. The amount of these partial calculation costs is included in the column matrix K_n. However, solving this task is not purely a pricing task but one of putting to good use its central idea, i.e., formulating a similar task with a given final use, but this modified method relates to the matrix of initial costs. In principle, this is a symmetric case to the pricing task. This symmetric case is defined by Relation (5):

$$K_n = [k_i] = \left[E_{n\times n} - A_{n\times n}{}^T\right]^{-1} \bullet Z_n \tag{5}$$

where the respective matrix represents:

K_n ... column matrix of unknown total costs.

Matrix of average costs is defined by Relation (6), where variable Q_i represents production volume:

$$P_n = [p_i] = \left[\frac{k_i}{Q_i}\right] \tag{6}$$

Relation (5) shows that we will work with unchanged initial costs and will calculate total partial costs for the respective cost objects, i.e., in our case assembly part classes. Here, the matrix of technical coefficients will be calculated from Relation (7). The technical coefficients must have the same values in this case, so the matrix marking will stay the same:

$$A_{n \times n} = \left[\frac{q_{ij}}{Q_i}\right] \tag{7}$$

where the respective elements represent:

q_{ij} ... the amount of pieces in the respective class.

4.4 *Calculating Calculation Costs for the Respective Assembly Part Classes Using the Symmetric Case*

The proposed input-output model scheme of the pricing task is shown in Fig. 3, where gray fields represent zero elements belonging to Quadrant I. The auxiliary matrix $T_{32 \times 9}$ that belongs to Quadrant I is given by the transposition of values in Table 7. The values for the column matrix of the final Y_{42} are acquired from Tables 4 and 5. The column matrix of production volume Q_{42} is given by a row sum of Quadrant I elements and the final use matrix. The initial costs are acquired from the records of the respective manufacturing units participating in the respective production phases.

The final relation is shown in Eq. (8) where the initial cost matrix already contains the values of the sums of initial costs for each class:

$$K_{42} = \left[E_{42\times42} - A_{42\times42}{}^{T}\right]^{-1} \bullet Z_{42} \tag{8}$$

To calculate the cost object costs, here an assembly group class, we use Relation (6). Finally, we can calculate the matrix of internal turnover for the main production activity of the company monetary units (9) that represents Quadrant I of the input-output model. For calculation matrix of final use, we apply Relation (10):

<table>
<tr><td></td><td></td><td colspan="3">Input</td><td></td><td></td></tr>
<tr><td></td><td></td><td>S</td><td>H</td><td>M</td><td>Y</td><td>Q</td></tr>
<tr><td rowspan="3">Output</td><td>S</td><td rowspan="3" colspan="2">$q_{ij} = 0$</td><td rowspan="2">$T_{32\times9} = [q_{ij+10}] \geq 0$</td><td>$y_i = 0$</td><td rowspan="3">$Q_i = \sum_{j}^{42} q_{ij} + y_i$</td></tr>
<tr><td>H</td><td rowspan="2">$y_{i+3} \geq 0$</td></tr>
<tr><td>M</td><td></td></tr>
<tr><td></td><td>Initial Costs</td><td colspan="2">Production Phase I</td><td>Production Phase II</td><td></td><td></td></tr>
</table>

Fig. 3 The scheme of the input-output model of the symmetric case

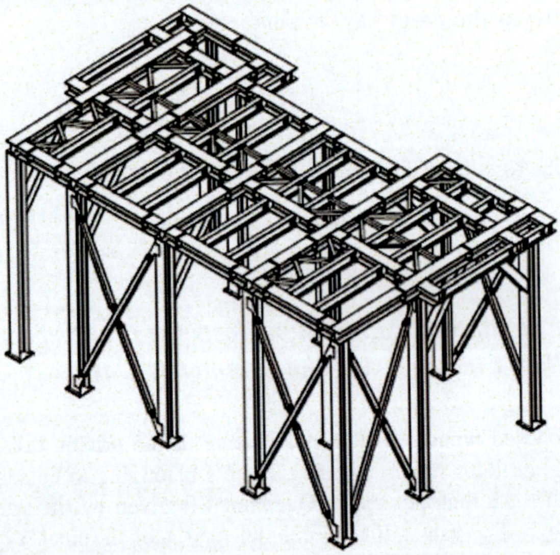

Fig. 4 Scheme of the steel construction

$$IT_{42\times42} = [p_i \bullet q_{ij}] \tag{9}$$

$$FU_{42} = [p_i \bullet y_i] \tag{10}$$

Final scheme of the steel construction is shown in Fig. 4.

5 Conclusion

The output of this work is mainly the testing of a proposed method published in an older article [6]. Testing was performed on a notional steel construction design within the requirements of the company Jetti. The basic benefit of this work is that the scope of the input-output model is affected by the quantity of components and assembly groups in the construction. This quantity is also reflected in the quantity of aforementioned classes that the proposed input-output model uses. Therefore, in the next phases of research, more focus will be given to decreasing the quantity of such classes in the respective class groups. Such decrease should have a positive impact on the actual production planning. To verify this claim, another notional steel construction design that respects the results of this testing will have to be proposed. Further work should be dedicated to making such comparison in order to confirm these conclusions.

After final discussions with the company mentor, it seems that such proposed methodology could become a common tool for cost management—not only in our company. However, it should be pointed out that the methodology is still being developed. Future research goals include completing a technological analysis that relates to the calculation of initial costs by modern cost management methods, in particular ABM/C and perhaps other stochastic tools of production activity management, e.g., PERT. The idea of this conceptual methodology could be extended in the future to include other main production activities, especially the custom manufacturing of high-temperature furnaces. However, there is also the possibility of using the power of input-output models in secondary production activities, especially in cooperation among manufacturing units, which is becoming increasingly important.

This work had been supported by the ČVUT Student Grant Competition No. SGS20/163/OHK2/3T/12.

References

1. J. Dubihlela, R. Rundora, Employee training, managerial commitment and the implementation of activity based costing; impact on performance of SMEs. Int. Bus. Econ. Res. J. **13**(1), 27 (2013). https://doi.org/10.19030/iber.v13i1.8354
2. A. Foroughi, M. Kocakulah, A. Stott, L. Manyoky, Activity-based costing: helping small and medium-sized firms achieve a competitive edge in the global marketplace. Res. Econ. Manag. **2**(5), 150 (2017). https://doi.org/10.22158/rem.v2n5p150
3. I. Mahal, A. Hossain, Activity-Based Costing (ABC)—an effective tool for better management. Res. J. Finance Account. **6**(4), 66–73 (2015)
4. B. Popesko, P. Novak, ACTIVITY-BASED COSTING APLICATIONS IN THE CZECH REPUBLIC. Lex Et Scientia. **XV**(1) (2008)
5. Jetti, a. s., https://www.jetti.cz. Accessed 20 Nov 2020
6. M. Machka, J. Suchomel, Návrh metodiky zpracování zakázky pro ocelovou konstrukci. Grant J. **9**(2), 100–103 (2020)

7. K. Janovská, M. Lampa, I. Vozňáková, Ekonomicko—matematické metody strukturální analýzy jako jeden z významných nástrojů optimalizace ekonomického řízení hutního podniku. Hutnické listy. **2010**(2), 83–88 (2010)
8. K. Janovská, I. Vozňáková, L. Švajdová, *The Verification of Applicability of Economical-Mathematics Methods of Structural Analyses as a Tool for Optimising Economic Proceedings of Metallurgical Enterprise*, in METAL 2010 (Tanger, Ostrava, 2010), p. 5. ISBN 978-80-87294-15-4
9. H.D. Kurz, N. Salvadori, 'Classical' roots of input-output analysis: a short account of its long prehistory. Econ. Syst. Res. **12**(2), 153–179 (2000)
10. W. Leontief, *Input-Output-Economics*. 2 (Oxford University Press, New York, 1986). ISBN 0-19-503525-9
11. B. Popesko, Š. Papadaki, *Moderní metody řízení nákladů: jak dosáhnout efektivního vynakládání nákladů a jejich snížení*. 2 (Grada Publishing, Praha, 2016). ISBN 978-80-247-5773-5
12. R.E. Miller, P.D. Blair, *Input-Output Analysis: Foundations and Extensions*, 2nd edn. (Cambridge University Press, New York, 2009)

Possible Methods of Valuing Startups

Jaroslav Hura, Samer Al-Rabeei, Peter Korba, Michal Hovanec, and Patrik Šváb

1 Introduction

Today, people with a startup concept can meet very often, and their definitions and views are very different. EPE and VCA define a startup as a company that is in the process of starting a business or shortly after starting a business but is not yet achieving sales [1].

Wells and Jeng, who work at Harvard Business School, consider a startup company as one that is moving from the stage of a business idea and preparing for the production, marketing, and sale of the product itself. The very first stage of a company's life cycle can be considered a business idea. At this stage, the founders have not yet taken any steps to implement it. Later, when they start developing a product or analyzing the market, they move to the so-called seed phase. Only then, at a time when they are starting a company, preparing production, marketing, and simply preparing the entire business model, are they moving to a stage where they can be described as startups. In Slovakia, the term startup is often referred to (especially in the media) as a business idea or a company in the pre-start (seed) phase [2, 3].

Eric Ries, a successful entrepreneur from Silicon Valley, emphasizes that startups are destined to deliver a new product or a new service to target customers in conditions of extreme uncertainty [2, 4].

Steve Blank states that the startup is looking for a repeatable and so-called scalable business model. This model is based on the potential to achieve significant

J. Hura · S. Al-Rabeei (✉) · P. Korba · M. Hovanec · P. Šváb
Department of Aerospace Engineering, Faculty of Aeronautics, Technical University of Kosice, Košice, Slovakia
e-mail: Jaroslav.hura@student.tuke.sk; Samer.abdo@tuke.sk; Peter.korba@tuke.sk; Michal.hovanec@tuke.sk; patrik.svab@tuke.sk

L. Knapčíková, D. Peraković (eds.), *6th EAI International Conference on Management of Manufacturing Systems*, EAI/Springer Innovations in Communication and Computing, https://doi.org/10.1007/978-3-030-96314-9_18

revenue growth without significantly increasing costs. A simple example is a mobile application—whether you sell 100 applications or a million, your costs are virtually the same, but sales (and profits) are significantly higher. The scalable business model is therefore a key prerequisite for achieving rapid growth [2, 5].

The startup in our conditions will practically not do without international (or global) ambition—that is, its goal should not only be Slovak, but, e.g., also a Pan-European or global market [2, 6].

2 Research and Finding

In the current situation associated with the pandemic of a new type of coronavirus SARS-CoV-2, startups are all the more important for economic growth because they create new jobs and introduce the latest technologies. Startups are also used by large technology companies, which use them to solve various tasks, and at the same time, these companies represent the necessary investments for startups [7].

But every viable startup must provide such goods and services to its customers in order to stay on the market and grow further. Such a successful startup will start to create jobs, thus reducing the unemployment rate in the economy.

The result of the pandemic of a new type of coronavirus SARS-CoV-2 is the beginning of an economic crisis that has affected individual economies. The rapid rise in unemployment in major consumer markets will lead to an almost immediate reduction in consumer spending. At present, a decrease of up to 50% is expected in some consumer categories. Ultimately, business expenses will decrease in all categories at the discretion of individual firms. As a result, companies with high growth potential, startups, and small and medium-sized enterprises, and thus the entire national economies of the world, will be quite significantly affected [7, 8].

Unlike startups in the startup phase, a functioning business with a history will have other hard facts and figures. The development of revenues and financial records facilitates the calculation of the value of the company.

One of the most popular startup valuation methods used by an angel investor points out that it is first necessary to determine the average valuation of startups in a given market area. Then, according to Forbes, it is possible to determine how startups compare to others in the same region by evaluating the following factors:

2.1 *Management Team Strength (0–30%)*

Founding team—The value will vary dramatically depending on the background and experience of the founding team. Jeff Bezos or Mark Zuckerberg could put a 10% stake in a new technology startup worth a nine-digit investment, while computer-illiterate friend Joe may be able to control only a few hundred dollars for the same stake.

2.2 *Opportunity Size (0–25%)*

Market size—You need to be interested in a pilot product. The larger the potential market, the better, especially if there are potential customers who are ready to buy.

Expected returns in the near future—If there are only a few potential customers who are not ready to buy, then the valuation will not be so high.

The finished product will still cost something, but ideally it will have enough power with 50–100 customers for investors to see the potential for returns in a short time. It is also important to note that 1 lead for $50,000 products is riskier than 10 leads for $5000 products.

2.3 *Product/Technology (0–15%) and Competitive Environment (0–10%)*

Competition—Entering a market full of competition is a risk and, as a result, the price will fall. In the case of a startup entering the open market space without competition, it can get a much higher investment.

2.4 *Marketing/Sales Channels/Partnerships (0–10%)*

Growth and engagement—Ideally, it is necessary to prove that the user base is growing and that people are engaged. If the startup offers the application and has 100,000 sporadic users, it is worth less than 20,000 loyal fans who use it every day. The shrinking user base is also a red flag that needs to be addressed quickly if investors need to be attracted.

2.5 *Need for Additional Investment (0–5%) and Others (0–5%)*

Like many methods, evaluating these factors is a very subjective process. It is important to remember that scalability and team are the main interest. As Payne states: "When building a business, team quality is crucial to success. A great team will fix early product errors, but on the contrary, this is not true" [9].

3 Methodology

3.1 Risks and Uncertainty in Business

Clearly, startups have been a current trend, especially in the IT-related industry, such as software development and application business. Even if a startup has a high failure rate, a market channel for an online business can generate revenue through technology. What seems to be a quick and easy way to start a business is actually trading with risk and uncertainty. The entrepreneur should therefore learn the successful factors of starting a business. Therefore, a good startup must be able to "grow fast" [4].

In business, some risk is necessary. To be successful, entrepreneurs need to learn to accept risks that cannot be avoided and to mitigate risks that can be managed. The best guarantee against risk is a good core team, a business-friendly government, diversified marketing channels, and prudent financial management.

3.2 Business Success

Not only in a market economy, but the goal of each individual or business entity is to satisfy their needs and maximize the benefits of each of their inputs. The company's performance is a criterion of financial decision-making and ensures the survival and competitiveness of the company in the market. One way to define a company's performance is the company's ability to achieve the desired business results in the form of outputs that are consistent with the company's goals, expressed in measurable units [4].

3.3 Measuring Business Performance

However, striving for the highest possible profit is also associated with higher risk and financial instability. Such a traditional approach to measuring performance and its use in financial decision-making in the company has been replaced in recent years by additional so-called modern approaches to performance measurement, which take into account several factors.

4 Result and Discussion

Valuing before starting a business without income can be a very difficult endeavor. Many things need to be taken into account, from the management team and market trends to product demand and related marketing risks.

After evaluating everything, even with the most effective pricing formula for starting a business before making a return, the best thing remains is that we can only hope, because it is only an estimate [10].

Therefore, it is important to list several ways you can value a startup without any revenue.

Initial valuation is the process of calculating the value of a startup company. Startup valuation methods are particularly important as they are usually applied to startups that are currently in the pre-income stage [10].

Most often, startups are valued only after they start operating, which means that the founders will not get as much as they expected, while investors pay more than they originally wanted to invest [9].

For this reason, it is necessary to get acquainted with the popular methods of valuing startups, based on which it is possible to better understand how to evaluate a startup company without returns and at the same time it will be possible to negotiate a better agreement with investors before achieving returns. Among the most used valuation strategies we can mention the following:

4.1 *Method No. 1: Berkus Method*

1. Quality of the idea (basic value)	300,000$
2. Existence of a prototype (technology)	500,000$
3. Team quality (value of the founding team)	300,000$
4. Strategic partnerships (successful market entry)	200,000$
5. Introduction and product sales	100,000$
Total value (max 2 million USD)	**1,400,000$**

Angel investor Dave Berkus thinks that investors should be able to imagine that the company will outperform $20 million in 5 years [9].

Table 1 Valuation of startups

Criterion	Weight	Target company	Factor
Team	30%	X	= 0.3*x
Size of opportunities	25%	X	= 0.25*x
Product/technology	15%	X	= 0.15*x
Competitive envir.	10%	X	= 0.10*x
Sales/marketing	10%	X	= 0.10*x
The need for more funding	5%	X	= 0.05*x
Others	5%	X	= 0.05*x
Total			The sum of all factors

His method evaluates five critical aspects of a startup:

- **Concept**: the product offers core value with acceptable risk.
- **Prototype**: this reduces technological risk.
- **Quality management**: if it does not already have it, then whether the startup has a plan to obtain quality management.
- **Mergers**: the existence of strategic relationships that reduce competitive risks in the market.
- **Startup plan**: the existence of a sales plan and preparation for the launch of the product (this does not apply to all startups before income).

Each aspect is assigned a rating of up to $500,000, which means that the highest possible award is $2.5 million. The Berkus method is a simple estimate that is often used in startup technology companies. It is a useful way of determining value, but as it does not take the market into account, it may not offer the extent to which people desire [10].

4.2 *Method No. 2: Scorecard Valuation Method*

This is one of the most popular business valuation methods used by angel investors (Table 1). It is also known as the Bill Payne valuation method and works when comparing a startup with others that already work and in which it has already been invested.

4.3 *Method No. 3: Venture Capital Method*

The risk capital method consists of two steps using several pre-investment valuation formulas:

Table 2 Evaluation expected startups

Expected value at the departure of the investor	Target return on investment	Valuation after investment				
100 mil. $	/ 20x	= 5 mil. $				
			invest-ment	Valuation after expected evaluation		
		5 mil. $	- 1 mil. $	= 4 mil. $		
					Expected appreciation in the amount of 30%	The current value of the startup
				4 mil. $	70%	= <u>2,8 mil. $</u>

1. It is first necessary to calculate the final value of the company in the year the investor leaves the startup.
2. Subsequently, it is necessary to monitor the expected return on investment, the value of the investment, and the current value of the startup (Table 2).

The terminal value of a startup is the expected value when launched on a specific date in the future, while the year of departure is the year in which the investor obtained the expected return on the investment. At the same time, with this method, it is necessary to know the ratio of the share price to its returns in the sector in which the startup operates. For example, a stock is valued at 3 × 1$.

The variables required, return on investment and amount of investment, are needed to calculate the present value. The current value of the startup can be calculated as the final value/return on investment—the value of the investment.

In a hypothetical example, if the investor requires a return on investment within 10 years with an investment value of 1 million $, present value = 20 million $/10–1 million $ = $ 1 million.

Using this method, it is possible to derive the current valuation before the actual start of the sale at 1 million $. With an investment of 1 million $ and the expected growth, revenues from the industry could have the value of 20 million $.

4.4 *Method No. 4: The Method of the Total of Risk Factors*

Using the method of the total of risk factors, the startup value of a startup will increase by $250,000 for each +1 or by $500,000 for each +2. Conversely, the valuation of a startup will decrease by $250,000 for each point −1 and $500,000 for each −2 points (Table 3).

This method is very useful in examining the risks that need to be managed in order for a startup to be successful [9].

Table 3 Startup valuation using the sum of risk factors

Original value				$1,500,000
1	Management	Very small	$500,000	$2,000,000
2	Stage of business	Normally		
3	Legislative and political risk	Normally		
4	Production risk	Normally		
5	Business and marketing risk	Normally		
6	Financing and capital risk	Normally		
7	Competition	Very high	–$500,000	$1,500,000
8	Technological risk	Small	$250,000	$1,750,000
9	Risk of litigation	Very small	$500,000	$2,250,000
10	International risk	Normally		
11	Reputation risk	Very small	$500,000	$2,750,000
12	Potential lucrative risk of leaving	Normally		

4.5 Method No. 5: Method of Market Valuation of Human Capital

Valuing a startup before starting operations is not an easy task for the investor, because most of them do not know the value of intangible assets. The potential investor should calculate the value of the ideas, know-how, and human potential of the team.

Projects can be awarded in two ways:

- Get to know the team and their expertise, evaluation of the people who develop the project
- Or perform a purely mathematical valuation based on the size of the market.

If the investor obtains at least rough estimates, he/she can easily estimate the potential of the startup, and thus the future profits hidden in today's valuation [9, 11].

4.6 Method No. 6: Asset-Based Valuation Method (Book Value Valuation)

The simplest valuation of a startup without income is an asset-based valuation, because it offers a reliable valuation of the real value of the startup company.

For this method, it is necessary to quantify:

- The initial cost of the startup company's assets offset by depreciation and other costs.

Table 4 Valuation of assets before the start of business

Stage	Investors	Valuation
Concept/business plan	Me or friends and family	250,000$–1,000,000$
Developed technology	Angels, initial funds	1,000,000$–5,000,000$
Start/early customer response	Initial funds, other funds	5,000,000$–15,000,000$
Scaling (without cash flow)	Other funds	15 million $–30 million $
Fast and massive expansion (cash flow)	Public companies, strategic investor	100 million $–1 billion $

- The total value of tangible assets is added to the amount in the balance sheet on the asset side, and also to cash and receivables.
- Any outstanding debts or costs are deducted from the total amount to obtain a value based on the assets of the startup company.

The problem with this method is that it takes into account the current state before starting a business and does not take into account how it will be in the future. Investors are more interested in the type of option, and therefore, as asset valuation does not take this into account, this method has some limitations [9].

In this method, assets are valued before the start of the business. No savvy investor would invest more than the market value of assets, so it is useful to know this value when looking for investors. For example, a technology startup can quantify the cost of developing prototypes, the cost of patent protection, and the cost of research and development before the actual start of operation (Table 4).

Unfortunately, this method does not take into account future potential, intangible assets such as brand value, or current market trends.

As this is generally an objective approach, it is, therefore, best to use this method as an estimate of the initial valuation of the startup before achieving income.

5 Discussion

Key factors to consider during valuation before starting a business before achieving revenue are given below. The true story of the company can be obtained by looking at the following:

1. Number of users—it is necessary to prove to exist customers. The more the better.
2. Marketing effectiveness—it is important to point out the ability to attract high-value customers at relatively low acquisition costs while gaining investors' attention to the possibility of making a profit.
3. Growth rate—to point to an already functioning business that has grown with a small budget, because many investors will see the potential for growth when they have some financial security [9, 12].

There is a link between these three concepts, as a strong marketing strategy will lead to impressive growth. If this happens, the number of users will increase. By providing evidence that there is a viable and scalable business idea, it will support the very startup of the business itself [9, 13, 14].

The value of the founding team cannot be forgotten either. Investors in a startup business want to be sure that they support a team that is destined to succeed. Consider this:

- Proven experience—if the team includes people with previous success in other startup companies, it will be more attractive than a startup full of inexperienced newcomers.
- Diversity of skills—ideally, the startup team will have a mix of experts whose skills complement each other. An excellent programmer cannot do everything on his/her own, but if he/she joins a marketing expert, the startup is worth it.
- Commitment—having great people is just part of the puzzle. These people need to have the time and determination to make sure that the startup is up and running. The team of part-time employees will not be attractive [9, 14].

No matter what formula is used to value expected returns, the prototype is a game-changing add-on. The ability to show investors, who are investing in a startup business, the functional model of the product offered not only proves the team's tenacity and vision to implement ideas, but also moves the business much closer to the launch date.

A minimum viable product or a startup that is expected to adapt to the market should be able to attract investments ranging from 500,000$ to 1,500,000$. A working prototype could gain even more if the company is examined by the stage valuation method used by many venture capitalists and angel investors. This can lead to investing 2–5 million dollars [9, 11].

Valuing a startup before it starts and earns money seems quite complicated, but fortunately, it is possible to draw on the experience and wisdom of other entrepreneurs, angel investors, and venture capital investors [9, 12].

It is important to note that there are several pitfalls that can lead to valuation errors when looking for a way to value a startup before making a profit.

The most serious ones include the following:

- It is never possible to assume that the valuation is permanent—the startup will cost only as much as investors are willing to invest. The startup owner does not have to agree with every award he/she receives for the business. It is important to realize that no rating, high or low, is ever permanent or even correct.
- It is never possible to assume that valuation is direct—there is almost nothing direct in business. In the event that the startup also obtains the expected investment, it is best to discuss all things in detail once again with potential investors, to make sure that investors know everything and are identified with the next steps [9, 15, 16].

6 Conclusion

Today's society is facing the challenge of overcoming the pandemic of a new type of coronavirus SARS-CoV-2. As a result of this pandemic, the global economy, as well as the economies of individual countries, will enter a recession. According to the International Monetary Fund, the global economy could recover as early as 2021, but important measures will need to be taken to achieve this recovery, leading to sustainable growth in national economies. As mentioned in this chapter, supporting startups is very important because startups create new jobs, bring new innovative technologies, and thus contribute to economic growth. It follows that there is not much time to take and implement effective measures in real life, but entrepreneurship is one of the key elements of a market economy and a way to increase a country's economic growth.

The actual valuation of startups can be very challenging. Many factors need to be taken into account, from the management team and the situation and trends in the markets to the demand for the product itself and, of course, the marketing risks. After evaluating all this, it is necessary to take into account that it is still only an estimate and the reality may be different.

References

1. Corporate Finance Institute, What are Startup Valuation Methods? [online]. 2020, [cit. 2021-02-15]. https://corporatefinanceinstitute.com/resources/knowledge/valuation/startup-valuation-methods/
2. L. Šrenkel, Čo je to start up? [online]. 2015 [cit. 2021-02-15]. https://www.podnikajte.sk/priprava-na-start/co-je-startup
3. Businessmap Ltd, Startup project management [online]. 2020, [cit. 2021-02-15]. https://flow-e.com/startup-project-management/
4. Digital Edenz, Why Startups are important for economic growth of a nation [online]. 2018, [cit. 2021-02-15]. https://digitaledenz.com/why-startups-are-important-for-economic-growth-of-a-nation/
5. D. Dluhošová, *Finanční řízení a rozhodování podniku: analýza, investování, oceňování, riziko, flexibilita*. 3. upr. vyd. (Ekopress, Praha, 2010). ISBN 978-80-86929-68-2
6. J. E. Pinto (ed.), *Equity Asset Valuation*, vol xx, 2nd edn. (Wiley, Hoboken, N.J, 2010), p. 441. ISBN 9780470571439
7. D. Herman, Jumpstarting Customer Demand Starts with Government [online]. 2020, [cit. 2021-02-15]. https://startupgenome.com/blog/jumpstarting-customer-demand-starts-with-government
8. G. Johnson, K. Scholes, *Cesty k úspěšnému podniku* (Computer Press, Praha, 2000). ISBN 80-7226-220-3
9. R. Richards, How to Value a Startup Company with No Revenue [online]. 2019 [cit. 2021-02-15]. https://masschallenge.org/article/how-to-value-a-startup-company-with-no-revenue
10. E. McGowan, 10 Real-World Startup Valuation Methods [online]. 2018 [cit. 2021-02-15]. https://www.startups.com/library/expert-advice/startup-valuation-methods
11. A. Riani, 5 Decisions That Will Increase Your Chances of Building a Successful Startup [online]. 2020 [cit. 2021-02-15]. https://www.forbes.com/sites/abdoriani/2020/09/19/5-decisions-that-will-increase-your-chances-of-building-a-successful-startup/?sh=29d9e3af21a3

12. Startup Genome, The Global Startup Ecosystem Report 2020 (GSER 2020) [online]. 2020, [cit. 2021-02-15]. https://startupgenome.com/
13. StartupDecisions.com.sg, Managing Startup Risks—An Entrepreneur's Guide [online]. 2019, [cit. 2021-02-15]. https://www.startupdecisions.com.sg/startups/launch-and-growth/startup-risk-management/
14. S. Nasser, 9 methods of startup valuation explained [online]. 2017 [cit. 2021-02-15]. https://www.techinasia.com/talk/9-method-startup-valuation
15. A. Wangai, U. Kale, S. Kinzhikeyev, An application of impact calculation method in transport. Transport **35**(4), 435–444 (2020). https://doi.org/10.3846/transport.2020.13909
16. M. Pinosova, M. Andrejiova, E. Lumnitzer, Synergistic effect of risk factors and work environmental quality. Quality Access Success **19**(165), 154–159 (2018)

Index

L. Knapčíková, D. Peraković (eds.), *6th EAI International Conference on Management of Manufacturing Systems*, EAI/Springer Innovations in Communication and Computing, https://doi.org/10.1007/978-3-030-96314-9